GENETICS

FOR

DOG BREEDERS

SECOND EDITION

Titles of related interest

ROBINSON
Genetics for Cat Breeders, 2nd edition.

EDNEY
The Waltham Book of Dog & Cat Nutrition, 2nd edition.

LANE
Jones's Animal Nursing, 5th edition.

EMILY & PENMAN
Handbook of Small Animal Dentistry

ANDERSON & EDNEY
Practical Animal Handling

VERMA & BABU
Human Chromosomes

GENETICS

FOR

DOG BREEDERS

SECOND EDITION

by

ROY ROBINSON F.I.Biol.
St. Stephens Road Nursery,
Ealing, London W13 8HB

PERGAMON PRESS
Member of Maxwell Macmillan Pergamon Publishing Corporation
OXFORD · NEW YORK · BEIJING · FRANKFURT
SÃO PAULO · SYDNEY · TOKYO · TORONTO

U.K.	Pergamon Press plc, Headington Hill Hall, Oxford OX3 0BW, England
U.S.A.	Pergamon Press, Inc., Maxwell House, Fairview Park, Elmsford, New York 10523, U.S.A.
PEOPLE'S REPUBLIC OF CHINA	Pergamon Press, Room 4037, Qianmen Hotel, Beijing, People's Republic of China
FEDERAL REPUBLIC OF GERMANY	Pergamon Press GmbH, Hammerweg 6, D-6242 Kronberg, Federal Republic of Germany
BRAZIL	Pergamon Editora Ltda, Rua Eça de Queiros, 346, CEP 04011, Paraiso, São Paulo, Brazil
AUSTRALIA	Pergamon Press Australia Pty Ltd., P.O. Box 544, Potts Point, N.S.W. 2011, Australia
JAPAN	Pergamon Press, 5th Floor, Matsuoka Central Building, 1-7-1 Nishishinjuku, Shinjuku-ku, Tokyo 160, Japan
CANADA	Pergamon Press Canada Ltd., Suite No. 271, 253 College Street, Toronto, Ontario, Canada M5T 1R5

First edition 1982

Reprinted 1989

Second edition 1990

Library of Congress Cataloging in Publication Data
Robinson, Roy.
Genetics for dog breeders/by Roy Robinson—2nd ed.
p. cm.
Includes bibliographical references (p.
1. Dogs—Breeding. 2. Dogs—Genetics. I. Title.
SF427.2.R63 1990
636.7'0821—dc20 90-6794

British Library Cataloguing in Publication Data
Robinson, Roy, *1922-*
Genetics for dog breeders. — 2nd ed.
1. Livestock: Dogs. Breeding. Genetic factors
I. Title
636.70821

ISBN 0–08–037492–1

Printed in Great Britain by BPCC Wheatons Ltd, Exeter

Preface to the Second Edition

THE LAST decade has seen a gratifying increase of interest in the genetics of dogs. This has ensured that a number of outstanding problems of breed genotypes have been resolved but, alas, others remain as enigmatic as ever. Not so fortunate is the increase over the same period in the number of anomalies which have been reported to be genetic in origin. This is due in part to a greater awareness that certain types of anomaly could be genetic and in part to commendable attempts to secure breeding data to confirm the supposition.

The increase in knowledge has meant that a new edition was called for to incorporate the findings. Chapters VI to VIII have been extensively rewritten, especially Chapter VIII. The latter continues to give descriptive summaries of anomalies which should assist the serious breeder to identify a suspected malady and to ascertain if it is likely to be genetic. In response to requests, the summary accounts have been substantially documented for the benefit of those (especially veterinarians) who may wish to consult the original report(s). In this respect, the chapter constitutes a useful reference source for identifying known genetic anomalies.

December 1989 ROY ROBINSON

Preface to the First Edition

THE WRITING of this book stems from the belief that the continued advancement of dog breeding will depend upon appreciation of modern trends in animal breeding. It should be acknowledged that the world of animal breeding is moving steadily away from "rule-of-thumb" methods in favour of a more balanced programme of selection and mild inbreeding. The thoughtful breeder would be wise to ponder on the fact that once his or her dogs are provided with a good home, are properly nourished and under expert veterinary supervision (when necessary), the only avenue left for substantial improvement resides in the science and art of breeding.

The early chapters present an outline of the basic principles of heredity. Some theory had to be considered, but this has been largely relegated to second place in preference to the more practical aspects. This is followed by a discussion of the many facets of selection, inbreeding, line breeding and creation of strains and blood lines. All of these are grist to the mill of genetics and should be considered as an essential backcloth to the art of skilful breeding.

Later chapters deal in some detail with an account of colour and coat inheritance, with particular attention to the fundamental types which often cut right across breeds as such. This latter aspect demonstrates the underlying unity of dog breeds. The topic is of considerable interest in itself, in addition to contributing to an understanding of breed evolution.

Alas, no book is complete without a discussion of the various anomalies which can afflict the dog. Fortunately, the great majority are rarely encountered. The publicity given to a few of the more common and serious (in part because they are common!) can obscure this fact. The conscientious dog breeder should be aware of those defects and ailments which are primarily genetic in nature. He or she should be aware of the clinical signs and have an idea of the steps to be taken to combat them. It is hoped that this chapter will be of service to veterinary surgeons in the competent pursuit of their profession.

ROY ROBINSON

Contents

1. Introduction

A FEW initial words may not be amiss upon the sort of information a dog breeder might be expected to find in a book devoted to canine genetics. Three aspects will no doubt immediately spring to mind.

Firstly, there will be an account of heredity and variation as embodied in the ideas of modern genetics. What are the genes and how are these passed on from parent to offspring? What is meant by genetic ratios and how are these produced by quite simple biological processes? An elementary yet adequate discussion of these questions will be given, choosing examples from the dog to illustrate the principles involved.

Secondly, a detailed account of the known genetics of the dog, the different mutant genes which have been discovered and their effects upon the animal. The heredity of colour and coat types, especially how these have been incorporated into breeds. The extraordinary variation of skull shape, body conformation and length of leg bones will be discussed: all of which feature prominently in the concept of breed differences.

Thirdly, modern methods of animal improvement which can be pressed into service to further the aims of dog breeding. What constitutes selection, when and how to apply it? How to distinguish between important and not so important traits in developing a successful breeding programme. What constitutes inbreeding, close and moderate forms, how and when to employ it? Is it fundamentally beneficial or harmful? Many questions of this nature deserve to be posed and answered.

It must be stated at the onset that it is difficult to offer specific advice on many problems of animal breeding. Only the breeder has all of the facts at his or her fingertips. However, this does not mean that general guidance cannot be given. On the contrary, the underlying principles are similar for all kinds of livestock. Armed with this knowledge, a breeder may find that he or she can make more decisive judgements or act with greater confidence. In the last analysis, it is the breeder who has to make the decisions, acting upon shrewd assessments of the potential of his or her dogs, backed up by the latest scientific information.

Antecedents of the Dog

The dog has the distinction of being the first animal to be domesticated by man. This was a unique event and seemingly was brought about by rather special circumstances. To appreciate how or why this occurred, it is necessary to journey back in time some 15,000 years or longer to imagine what life must have been like for prehistoric man. For many hundreds of thousands of years man was a hunter-gatherer. He hunted game of all kinds and he gathered roots, fruits and cereals in season. As time elapsed, he gradually lost his overt bestiality and "modern man" emerged, still largely in the hunter-gathering stage, but now entering upon a more settled mode of life in the form of primitive agriculture.

It is almost certain that several factors, but two in particular, forced this change upon mankind; increasing population and cultural evolution. Man was becoming more adept at modifying the environment to his liking. The events leading up to the change were slow, but the changeover occurred relatively quickly once it had begun. One of the major agricultural civilizations emerged in the Middle East, spreading both northwards and eastwards. This is the area where most of the important economic mammals were first domesticated, the goat, sheep and cow.

It is equally necessary to delve back into the past and consider the early ancestors of the canid species of today. It seems generally accepted that the first real dog-like mammals were emerging during the Pliocene, approximately 5 million years ago. By the Pleistocene, some 3 million years later, a primitive wolf (*Canis etruscus*) had developed which could have been the progenitor of several extant species, among them the modern wolf (*Canis lupus*). In any event, the wolf emerged as a highly successful predator, formerly colonizing a vast area of the world, ranging from most of North America and the whole of Europe and Asia as far south as Arabia and India. There is some argument whether the first wolves originated in North America or Eurasia, but, since the two continents were connected by a land bridge across the Bering Straits at least once during the period, this is an academic point.

The enormous territorial range of the animal would imply that different sub-species could be involved. There are, indeed, differences of size and coat quality. The northern races are larger and possess a thicker coat (e.g. the timber wolf), in comparison with the somewhat more gracile southern sub-species. For some time, the red wolf (*Canis rufus*) of North America was thought to be a distinct species, but meticulous study reveal no differences other than body size. The form is at best a sub-species, but more probably a local race isolated in southern Texas.

The expanses of Eurasia offer a better chance of finding sub-species, and in fact several have been named. These have a taxonomic value, for reference purposes, but little biological meaning. The southern Asian or Indian

Wolf (*Canis lupus pallipes*) is smaller than its northern cousins and tends to resemble those pariah or mongrel dogs which have undergone only minimal selection by man.

The two other modern dog-like species are the jackal and coyote. Both occur at more southerly latitudes than the wolf, the jackal particularly, and are smaller animals. Of the two, the jackal is evidently the most ancient, as shown by the fact that several distinctive-looking species can be recognized. These occur regionally in Africa. It is debatable how early the jackal separated from the wolf, but that they have a common ancestry is indisputable. Jackal-like remains date from the late Pliocene, hence the separation is very old and of sufficient standing for species differences to have arisen. The most abundant jackal is the common or golden (*Canis aureus*), occurring throughout North Africa, Arabia, India and the Malay peninsular. Note the overlap of the jackal and wolf in Arabia and India. It is tempting to speculate that the jackal evolved in Africa, separated from the main stream of wolf evolution by the Mediterranean basin, and adapted to a warm climate. Once a mating barrier has evolved, the jackal could extend its range northwards without loss of its identity.

The position for the coyote (*Canis latrans*) is different. No connection exists between the jackal and the coyote, the latter occurring only in the New World. Only one species exists, and this appears to be of relatively recent origin. Hall (1978) considers that the coyote may have descended from the wolf about 500,000 years ago. This would make the coyote a relative newcomer onto the scene. Hall proposes that the coyote may have evolved from relict pockets of wolves left behind by the receding ice ages and which adapted to a warmer climate, becoming more and more distinct from the main wolf population.

Although most people would regard the coyote, jackal and wolf as "good" (i.e. distinct) species, the process of differentiation has not proceeded as far as one might expect. Not only are the morphological differences relatively small, but all these species will hybridize and produce fertile offspring. This occurs infrequently, if at all, in nature, since the speciation has proceeded sufficiently that mating barriers have come into existence, but can be induced in situations where individuals of opposite sex are artificially kept together. The mating barrier is an indication of separate species, while the production of fertile offspring is an indication that these species possess a high degree of affinity. The domestic dog will breed with each of the species without too much difficulty. In all cases it is largely a matter of rearing the intended parents together from young pups.

The almost total absence of hybrids in the wild does not mean that it cannot occur, given special circumstances. For example, during the late 1950s a large coyote-like canid was observed in the state of New England. Initially considered to be a coydog (hybrid of coyote and dog), it now seems more likely that they are coyote-wolf hybrids. Most of the animals are intermediate to the coyote and wolf, but some resemble the pure species. The hybrids were able to arise and flourish because of the near extermination of the wolf and coyote from the area. The natural antipathy of the two species was overcome due to disruption of the habitat and to extreme paucity of mates of the same species. Evidently, a hybrid population was born which expanded to fill the void (Hall, 1978).

The coydog does, as a point of interest, occur at a low frequency in many parts of the U.S.A. In fact, for a time it was wondered if a new race of dog was evolving to become a local nuisance. Most hybrids are due to a chance encounter of a roaming dog bitch with a coyote male. In general, however, there appears little chance of the coydog becoming numerous and a menace (Mengel, 1971). Although apparently vigorous and fertile, they are less fit than either the wolf or feral dog. It is interesting, nonetheless, that the hybrid will occur naturally. Only in very favourable circumstances would they become numerous.

One of the fundamental attributes of a species is the characteristics of its chromosomes, not merely the number but their sizes and shapes. Examination of the chromosomes of the wolf, jackal and coyote reveal that these are identical in each species. As a point of fact, they are identical to those of the dog (see later). This explains in large measure why the hybrids between the species are not only possible but fertile. It also shows, most emphatically, the close affinity between the various species and the dog.

From an evolutionary viewpoint, it may be argued that the coyote, jackal and wolf constitute a species complex of a singularly adaptable mammal. A complex in which it is convenient to recognize three primary species and a number of secondary and sub-species. Several species of jackals are indeed depicted. However, it would seem that the separation of the species is well underway, but has not attained full completion since it is still possible to obtain fertile hybrids.

There is little doubt that the dog belongs to the complex, as shown by the fact that it will produce fertile hybrids with the various species and the identical karyotype. Few truly wild dogs exist today (as distinct from feral dogs), but the dingo could be held up as an example of a wild dog. The dingo probably evolved from an Asiatic wolf as a result of close association with early man. Fortunately for the dingo, it was transported to Australia where it persisted as a consequence of two factors. The absence of a serious wild competitor and the continuation by the Australian aborigine of a way of life conducive to its survival. As such, the dingo is living proof of the adaptability of the wolf complex. In this case, the complex has exploited an ecological niche opened up in Australia by the relatively unchanging mode of life of the aborigines.

It is true that the dingo has been awarded species status (*Canis dingo*), but today few people doubt that in reality it is a dog. The reasoning of Epstein (1971) and the work of MacIntosh (1975) has clarified the situation. The discovery of an intact complete skeleton, dated 3000 years before present (BP), showed that the animal has scarcely changed in the intervening years. This is further shown by fragments of teeth and jaw bones subsequently dated as between 8500 and 8000 BP. This places the dingo within the early period of domestication. This concept is supported by its appearance and behaviour. The coat colour is light wolf grey, the tail is still bushy, straight and pendulous, not lightly haired, curved upwards or held erect as found in most modern dogs. The dingo will accept man as a social companion, but will not respond to obedience training. To clinch matters, so to speak, the dingo has a karyotype which is identical to that of the dog.

It may be added that none of the various foxes (particularly the red fox, *Vulpes vulpes*) belongs to the wolf complex, despite the close physical resemblance. The number of chromosomes for the foxes is only 34 or 38 (compared with 78 for the wolf) and they are totally different in shape. Hybridization between the fox and dog has been alleged from time to time, but this has been disputed. Indeed, most authorities are very doubtful whether such hybrids are possible. Even if they are possible, the great disparity of chromosome number would almost certainly mean that they would be sterile.

Domestication

There have been speculations how the dog became domesticated. Alas, many of these are inclined to be romantic, rather than factual, for the origin of the dog is probably prosaic in the extreme. It is difficult to believe that the domestication was the result of deliberate policy, mainly because it occurred too early in man's history. The idea of domesticating animals had not begun to take shape. Almost certainly, the ancestors of the dog came into contact with man as a scavenger around "kills" of game and rousting among camp-sites and refuse dumps. As settlements grew more permanent, so the ancestors would associate with man, his activities and probably his huts, as a source of food. They would clear the area of edible debris of all kinds. Conceivably they would be tolerated or even welcomed by the inhabitants for this purpose.

As Zeuner (1963) has expressed the situation: "It is, therefore, not impossible that young wolves and jackals which grew up in or around the temporary camps of Mesolithic hunters would quite naturally regard the men who provided part of their food supply as members of the pack, an association which the hunters would not have failed to turn to their advantage." To comprehend the implications of this conjecture, it is necessary to remember that man and dog are both socially orientated and group-forming. They will instinctively follow and respond to a leader. Young pups, if caught early enough will form firm attachments with man. It is a short step to imagine that the semi-tame wolves or jackals may have participated in the hunt by either flushing or cornering game. Their reward would be offal, which man rejects but the dog relishes. Later, the dog may have acquired another role, that of guard dog, giving warning of marauding beasts or the approach of strangers.

The precise origin of the dog is lost in the mists of time and will never be conclusively known. It is doubtful if it will ever be possible to be more exact than to give a continental region. Similarly, there are considerable problems in dating the period in which teeth, jaw and other bones began to take on the appearance of dogs, rather than that of wolf-like dogs. But, nonetheless, the careful archaeological work of the last few decades have turned up some fascinating finds. Evidence of dog-like creatures have been discovered as far afield as the Palegawra cave, Iraq (12,000 BP), Ein Mailaha, Israel (12,000 BP), Jaguar cave, Idaho (10,400 BP), Star Carr, England (9500 BP), Eastern Anatolia (9000 BP), Lake Baikal, Russia (9000 BP), Devil's law, Australia (8000 BP), Mount Burr, Australia (8000 BP), Sian, China (6800 BP), and Benton County, Missouri (5500 BP) (McMillan, 1970; Olsen and Olsen, 1977; Davis and Valla, 1978).

In two instances there are indications that the remains were buried in graves, one with a human. The act of burial implies compassion for the animal, and is a frequent practice of people of a later date who have domesticated the dog. All of the above dates should be considered as approximate, and they could underestimate the antiquity of the early dogs, since for the remains to be different from the wolf the processes of domestication must have been in action for many generations for the changes to arise. On the other hand, plus or minus a few hundred years would make little difference in view of the time-span involved. All in all, domestication of the dog almost certainly antedates, or at least accompanies, man's own "domestication", i.e. the rise of villages and agriculture which gave birth to modern civilization.

One of the dog burials is attributed to the Natufian, a semi-nomadic culture of the Levant which predates the early agriculturalists. It is doubtful if it will be possible to delve much further back in time. The Middle East is probably the most likely area for the domestication of the dog, if only for the fact that the civilization which spawned in the area also bequeathed to the world the domestic cat, cattle, goat, sheep and perhaps the pig. It is probable that the dog was domesticated over a wide area, especially as man was increasing in numbers and was beginning to have a significant impact upon the environment. As large game became less abundant, man and dog could become more interdependent.

To be brief, it is highly probable that the dog was first domesticated in south-west Asia (Epstein, 1971), whence it spread around the world. It was carried to Australia and other islands by the native aborigines. How it reached North America is more contentious. Early man crossing the Bering Straits could have taken the dog with him or the knowledge that the wolf could be tamed. Or, of course, the domestication could have been discovered independently, although this seems less probable.

There is still the question from which species did the dog descend? Presumably the coyote is ruled out (if merely because of geography), hence this leaves the wolf or jackal. At one time the jackal was considered to be an equal candidate to the wolf, but this viewpoint is waning. The most likely ancestor is the Asiatic or Indian wolf (*Canis lupus pallipes*), a race which is smaller than the large northern wolves, very adaptable and is highly organized socially (more so than the jackal, according to some authorities). The only dogs likely to have descended from the jackal are the native Pariah breeds of Africa, but even this is doubtful. If the jackal did feature in the evolution of African dogs, it was probably as an admixture rather than pure jackal.

According to Zeuner (1963), the number of finds of remains or prehistoric dogs increase steadily in numbers and diversity over the centuries. There is variation of size and skull shape. This, of course, indicates increasing domestication and probably controlled breeding. There is evidence that certain types of dogs are associated with specific cultures, as if these had been selectively bred. A small dog for hut dwellers, for example, or a large dog for sheep and cattle driving, and as guards. Zeuner notes that dogs would have been especially useful to man in this respect. He suggests that some of the primitive sheepdogs of today may trace back almost unchanged to the early dogs of some 3000 years ago or even earlier. This, of course, is debatable, but the principle involved is clear. Also, during this whole period, there would be selection for tractability and obedience. Unruly and vicious dogs would be eliminated in short order.

By the time the Egyptians began to leave their imprint in the form of tomb painting and various artefacts, it would seem that most of the major groups of working dogs had made their appearance. There were large or medium sized, stocky or slender-built animals. Even a dog with exceptionally short legs was illustrated. A massive mastiff-like dog is portrayed on a terracotta tablet from Mesopotamia. These large dogs also occurred in India (Zeuner, 1963; Epstein, 1971). There appears to be less evidence that small pet or toy dogs were numerous at that time. Possibly only the well-to-do possessed such animals. The Chinese and Tibetan people are credited with creating the very small decorative breeds, but at a later date when civilization had progressed above the subsistence level. The three most prevailing dogs appeared to be a medium-sized animal for herding sheep (particularly), a medium or large, long-legged animal for hunting and a large, very muscular animal for guarding livestock, barns and dwelling houses.

The great diversity of modern-day breeds arises from a variety of causes. Certain breeds tend to be very similar in form and behaviour, even to the extent of possessing much the same colour, and there is little genetic justification for uncritical acceptance of claims of independence. This prolification is a modern tendency and is motivated in part by a desire to possess a unique dog. Apart from this, however, there still remains considerable diversity. The primary reason is the very considerable period of selective breeding which has taken place, a period which has to be measured in terms of some 8000 to 10,000 years or longer; and seemingly a similar number of generations. During this period, the dog spread throughout the world, where it was exposed to two influences. One was the environment, the second man's influence as he endeavoured to mould the animal to fulfil different functions.

The environment would probably be the more important factor initially, since the dog was unlikely to be properly fed and housed. This would come later as the worth of the animal became widely appreciated and it began to assume some economic value. The dog had to adapt to man and his environment, yet be of such a constitution to be able to fend for itself, at least partly. The type of dog which could evolve under these conditions would steadily diverge from the wild ancestor in a number of characters, in appearance and in behaviour until it is scarcely recognized as a wild animal. The question of appearance is of importance for livestock breeders. They would wish to readily distinguish their dog from prowling wolves.

The outcome was a somewhat more gracile animal, a little smaller in size, with less formidable jaws and teeth. There would be some selection by man in this respect, perhaps performed subconsciously, but it has been proposed that the changes occurred because the dog had no need to kill prey on its own account. Thus, strong teeth are still required to masticate food obtained by scavenging or scrounging, but this is far removed from the type of jaw required to bring down prey which may be resisting.

At a subsequent stage, man himself began to subconsciously change the dog for better or worse. Social conditions were strongly influential in the matter. Sheep breeding was an important occupation in Eurasia and shepherds required a highly intelligent animal which would herd sheep (but not kill), be obedient and responsive to command. Hence the development of the many breeds of sheep dogs. On the other hand, guard dogs were required to have quite different attributes. These had to be robust, of fierce mien and suspicious of strangers. Yet another type was necessary for purposes of hunting, sometimes of a specialized nature. Some hunting dogs had to be very fleet of foot, so as to out-run the quarry, while others had merely to indicate the position of the quarry or to retrieve shot game. Finally, there is a general category of house dogs, some of which had to combine companionship with guard duties, while others were almost exclusively ornamental. Thus, miniature and toy breeds were developed to satisfy a need for the latter.

How did all of this diversity come into being? Foremost, of course, the dog had to possess the innate genetic variability. Little would have been possible without such variation, but this is not the whole story. The genetic potential still had to be exploited. The answer lies in the great number of generations of selective breeding to which the dog has been subjected and the uncountable millions of animals which have been bred. From these were winnowed an animal which is basically docile, tractable and faithful, yet with specialized qualities.

The most recent development has been the show dog. The utility breeds still thrive, of course, but the show dog has come to the front in the ordinary person's estimation. Many have a veneer of utilitarianism, despite being bred for exhibition purposes. Other breeds have been developed solely as show animals and to satisfy a demand for this type of dog. In a manner of speaking, this is the end of the evolutionary road for the dog. Such animals are totally dependent upon man for substance, mates and companionship. This comment also applies to the mongrel hound of towns and cities all over the world. But does man derive any benefit from this relationship? The answer must be yes, for he has evolved a trusting and faithful friend, especially for the elderly, to whom the dog can be a boon in a rapidly changing world.

Several writers on the dog have made much of the variation of size and conformation, even to the extent of giving certain representative forms formal Latin designations (Zeuner, 1963; Fiennes and Fiennes, 1968). These may delight the taxonomist and they have a certain reference value, but it is wise to be cautious in making inferences of remote ancestry. The variation merely implies that the wolf had considerable potential genetic variation which man was able to exploit. Many of the genealogies of breeds of dogs make specious, rather than convincing, reading. As one researcher (Peters, 1969) could write: "...investigations revealed that records compiled prior to the middle of the 19th century are so few, incomplete and inaccurate that a person can 'prove' almost anything he cares to regarding specific breed ancestry". This structure applies equally to scientific treatises, so that any writing of this nature should be approached with caution. It is unfortunate that so many authors have uncritically repeated the speculations of earlier writers. The earlier conjectures may be legitimate, *qua* conjectures, but through the passage of time these take on the mantle of authoritative facts which is not entirely warranted.

Just how much influence a local race of wolf or jackal may have had on the development of the dog is open to question. This would largely depend upon circumstances. Two examples will make the point. Eskimo dogs have to be hardy beasts to withstand the rigours of sub-arctic conditions, and the opinion has been offered that these dogs are descended from northern wolves or have been crossed with northern wolves from time to time. This is entirely feasible, but also is the alternative explanation that the harsh environment is shaping all sub-arctic dog breeds to appear like northern wolves in order to survive. Namely, a heavily built dog, with a thick coat, is the ideal conformation for breeds inhabiting the region.

In a similar manner, it could be suggested that several African breeds are descended from the jackal, either directly or indirectly, by crosses with the jackal in order to breed a dog more suitable to tropical conditions. This is entirely possible, but so is the idea that the environment would be shaping African breeds to assume jackal characteristics in order to survive. Thus, the argument could go around and around, with no clear-cut answer. Yet, why seek a yes or no answer? It is patent that the environment will be shaping the dog, whether consciously helped by man or not.

Another accomplishment of domestication is a pronounced increase in fertility. The increase in fecundity for the dog over that of the wolf is as remarkable as that for body size variation. The wolf and jackal are subject to seasonal breeding, normally littering in March to June for the former and January or February for the latter. The average litter is four pups, with a range of one to seven. Sexual maturity occurs at about 2 years for both species.

The dog, on the other hand, can reproduce at any time of the year, as soon as the bitch becomes sexually mature at about 1 year. Litters are born 7 or 8 months thereafter, instead of yearly, and are composed of between six to eight pups on the average. The only feature which is unchanged is the pregnancy period which is about 60 to 63 days for all species. It is interesting that the dingo and Basenji dog have a marked tendency to be seasonal breeders, littering about January to April. The dingo may have one to eight pups per litter with an average of four, while the Basenji has a similar range of one to eight pups, but a slightly higher average of about five. These dogs seemingly still retain some of the characteristics of the wolf.

Whatever may be the precise origin of the dog, there is no doubt that present-day dogs are a single entity. By this it is meant that dogs brought together from any part of the world will breed together and produce fertile offspring, even if a preliminary process of familiarization may be necessary in some cases. Even the dingo will breed with the dog, despite a preference to mate only with its own kind (MacIntosh, 1975). This mating preference is probably akin to, but more strongly expressed than, the discriminatory acceptance or rejection of mates shown by ordinary dogs. Thus, the dingo is indicating that it is different, but not that different, from the remainder of the dog population.

It is intriguing to consider if the long evolution of the dog has left behind what could be termed primitive breeds of the domestication process. The dingo has already been briefly mentioned in this respect, for it exhibits several features of what could be termed a first-generation domesticate. The tamed dingo appears to tolerate man rather than to welcome him, while the wild dingo is virtually a wild dog or at least a true feral dog. Certain areas of Australia provided a suitable environment for the dingo to thrive as a relict population which has never attained full domestication. Although generally similar to other native dogs, the dingo has unique characteristics which can be appreciated by experienced eyes (MacIntosh, 1975).

Another link with the past are the free-ranging dogs of the native tribes of Malaysia and Africa, of which the Basenji has been studied and may be regarded as typical in a general sense. The Basenji, similar to the wolf and dingo, is a seasonal breeder and is not as reproductively prolific as the modern dog. The breed may be depicted as a further step along the road to domestication. It seems to be more socially related to man than the dingo and to participate more in his affairs. Many of these ancient native breeds are in danger of extinction, due to the penetration of modern dogs into even the remotest villages, with consequential inter-breeding.

A final point may be made. No matter how diverse the many breeds of dog may appear for coat colour, fur texture or length, size, body proportions, stance or behaviour, all of the variations can be readily interpreted in terms of gene mutation. The various permutations of these characters to produce the many known breeds are the outcome of selective breeding by man, either as part of the process of domestication, its long aftermath, or the recent accentuation of differences in the modern show animal.

2.

Reproduction

THERE IS little point is discussing reproduction in the dog in detail, but a general summary will place the topic in perspective. Most breeders should be familiar with the overall picture. Almost all books on dogs devote a chapter or two to breeding, under such titles as the stud dog, the bitch in heat, whelping, and so on, the emphasis being upon the practical aspects. A few books have dealt with the subject at considerable length (Portman-Graham, 1975; White, 1978).

In the main, mating and parturition are normally straightforward events, but complications can arise. Pedigree dogs are valuable animals, but in terms of money and affection, and the general supervision of a vet would not be amiss. This is sound advice for the novice, for the experienced breeder quickly learns to detect signs that all is not well for a pregnant or lactating bitch, and will seek help accordingly. For the technically minded, an excellent review of canine reproduction is provided by Stabenfeldt and Shille (1977).

The young puppy is a playful and inquisitive creature, full of energy interspersed with bouts of restful sleep. Given reasonable care and attention, the little animal soon outgrows his puppy-like behaviour, becoming mature at any time between 6–12 months, depending upon breed and adult body size. Size is the most important determinant, the smaller breeds attaining sexual maturity, or puberty, earlier than large breeds. It is usual for the male to lag several weeks behind the female in this respect, as judged by eagerness to copulate and the presence of abundant sperm in the seminal fluids or semen. That age, rather than season, is the main factor in the onset of oestrous, or heat, is shown by the absence of peak months for births throughout the year. Few breeds give more than a mild indication of optimum months for mating.

Many breeders believe that bitches come into heat twice a year. This is a good rough and ready guide but is not strictly true. The interval between oestrous cycles is between 6–8 months on the average, but many bitches may come into heat much sooner or not until much later. The range may be as great as 5–11 months. Differences in body size do not seem to be of great importance, as might be expected, whereas breed is a factor. Bitches of certain breeds appear to have more regular oestrous cycles than others.

TABLE 1

Basic Data on Reproduction in the Dog

Feature	Average	Typical variation
Puberty, male	9 months	6–12 months
Puberty, female	10 months	7–13 months
Oestrous	8 days	6–10 days
Inter-oestrous period	6–8 months	5–11 months
Gestation period	62–63 days	59–66 days
Litter size, Small breeds	3–4 pups	1–5 pups
Medium breeds	6–7 pups	2–10 pups
Large breeds	7–8 pups	3–15 pups
Sex ratio (male:females)	103.4:100	—
Lactation period	6 weeks	5–8 weeks
Breeding season	Jan.–Dec.	—
Longevity	10–13 years	12–15 years

That is, the intervals between heats is less erratic. This is a feature which could be significantly controlled by strain. The oestrous interval is often extended by a few weeks in older females.

Immediately preceding oestrous, there is a "run-up" period, lasting from 7–12 days, during which the ovaries are preparing to release the eggs and the uterus is preparing to receive the eventually fertilized eggs. The lining of the uterus becomes engorged with blood and some escapes via the vagina. This is the so-called pre-oestrous bleeding, an external sign of the onset of oestrous. The actual heat, or receptivity of the bitch for the male, extends from the last day or two of the oestrous bleeding and for several days following. Ovulation, or the release of eggs, occurs at the cessation of bleeding, roughly at the mid-point of heat.

If mating is successful, the fertilized eggs are gently floated into the uterus and attach themselves to the delicate lining by about the 17th to 20th day. Previously the eggs have been actively dividing but remain essentially a jelly-mass of cells. Once attachment has occurred, however, a placenta is formed and embryonic development proceeds at a quickening pace. Pregnancy lasts approximately 63 days, with some variation in practice, due to the fact that the actual day of ovulation (and hence fertilization) may be unknown. The developing foetuses may be felt through the wall of the abdomen by careful palpation during the final stages of pregnancy. At parturition, the delivery of an individual pup is rapid, but the interval between successive pups may be as brief as 5 minutes to as long as

60 minutes. Table 2 lists average gestation periods for a number of breeds (Krhyzanowski *et al.*, 1975).

In the days leading up to the oestrous the bitch will usually show behaviour changes. Her appetite may increase (or decrease), there may be more frequent urination and heightened interest in males. The male will

TABLE 2

Gestation Period (days) for a Few Representative Breeds

Breed	Average	Typical variation
Boxer	63.5	59–68
Collie	62.4	59–66
Dachshund (Smooth)	62.5	59–67
Dobermann	62.8	58–68
Fox Terrier, Wire	62.6	59–67
German Shepherd Dog	62.1	58–65
Great Dane	62.6	60–65
Pekinese	61.4	57–65
Pointer, Smooth	62.3	59–65
Pointer, Wire	62.2	58–66
Poodle, Miniature	61.6	59–66
Poodle, Standard	61.5	58–66
Poodle, Toy	61.8	58–67
Spaniel, Cocker	62.4	59–66

correspondingly show increased interest in the bitch, particularly when she commences to present herself. This involves holding the tail to one side, raising her pelvic region and standing quietly. The act of mating or coitus, in the dog is more than usually complicated. The male will clasp the body of the bitch with his forelegs and mount the bitch from behind. This is acccompanied by rapid pelvic thrusting and stepping from one hind-leg to the other as soon as intromission is accomplished. However, the male is unable to withdraw until detumescence of the penis occurs. He will dismount, but remains in a "copulatory lock", tail to tail with the bitch, for as long as 30 minutes. The dogs should not be forcibly separated during this period, since severe damage to the external genitalia of both sexes may occur. The lock is due to vascular engorgement of the penile bulbus glandis and contraction of the vaginal muscles. Some bitches display marked preferences for certain males and/or discrimination against others. However, this is rarely an insurmountable problem and may be overcome by a period of acclimatization.

Generally speaking, litters are born thoughout the year, but with a small seasonal rhythm which leads to slightly more whelpings during May to

July and the least number during January to February. In contrast to the above, which is for breeds as a whole, a few breeds have a greater tendency to litter in the autumn, during the months of October to December. Again, the tendency is small and would only be apparent for a large number of whelpings (Tedor and Reif, 1978).

For most breeds, however, the seasonal tendency is so small that for all practical purposes it may be disregarded. However, there is a remarkable exception. Bitches of the Basenji breed show a marked tendency to have their litters during the months of November to January, with only an occasional litter in any other month. This behaviour is probably significant, in that it suggests that the Basenji has retained certain primitive features of reproduction although being a true dog in other respects. Scott *et al.* (1959) have shown that the tendency of the Basenji to have autumn oestrous cycles (hence autumn litters) is inherited and differs from the ordinary domestic dog by merely a few genes.

The number of pups in a litter, or litter size, may vary from 1 to 20 or more in very exceptional cases. The average litter over all breeds is in the region of 6 pups, but this figure is meaningless because litter size is strongly influenced by the size of the dam. Large breeds have larger average litters (7–8 pups per litter) than medium (6 to 7) and small breeds (3 to 4). These averages conceal to some extent the true situation, for almost all of the very large litters are born to large or medium-large breeds. Breed is a factor to consider, for some breeds are exceptionally fertile, mainly those in the medium-large size range. Some of the small breeds are relatively infertile, have consistently small litters. This is definitely an aspect which should be closely watched. One of the fundamental reasons for the association between big litters and body size is that large bitches possess larger ovaries than small bitches and this enables them to release more eggs at any one oestrous. They also possess larger uteri to accommodate many developing foetuses (Lyngset and Lyngset, 1970; Robinson, 1973). Within a breed, there is a notable tendency for an inverse relationship between number of foetuses carried and gestation period. Averaging over a large number of litters, the larger the number of foetuses, the shorter the gestation. The trend is most noticeable in, if not confined to, medium and large breeds with a wide spread of litter sizes (Robinson, 1990).

The average litter sizes of Table 3 may be underestimated because almost all of the information is extracted from stud book records and breeders may not be punctilious in reporting still-births. It is not easy to obtain reliable figures on this aspect, but Druckseis (1935) has given some extensive data for the German Shepherd Dog. The proportion of still-births was 2.3 per cent over all litters, or about 0.18 pups for an average litter of 8. The percentage is reasonably constant over all litters, except for large litters of 15 or more, when the mortality jumps to 7.5 per cent. Significantly more males than females were observed among the still-births.

Evidently, bitches with very large litters can be expected to have problems. It would be interesting to have information to indicate whether large or small breeds vary in the amount of still-born pups or if specific breeds have exceptionally high or low percentages. Over all breeds the percentage may not vary a great deal.

Another aspect upon which it is difficult to obtain figures is the percentage of foetal atrophy or deaths between implantation and birth. This occurs regularly in most mammals and arises from a variety of causes. A proportion of the deaths are usually genetic, where a foetus has inherited

TABLE 3

Sizes of Litter for Various Dog Breeds

Breed	Average	Typical variation
Airedale Terrier	7.6	2–6
Basenji	5.5	1–8
Beagle	5.6	1–9
Bedlington Terrier	5.6	1–11
Bloodhound	10.1	2–10
Boston Terrier	3.6	1–7
Boxer	6.4	1–12
Bull Terrier	5.9	2–10
Bulldog	6.2	1–7
Cairn Terrier	3.6	1–7
Chihuahua	3.4	1–6
Chow Chow	4.6	2–9
Collie	7.9	2–13
Dachshund, Long Hair	3.1	1–6
Dachshund, Smooth	4.8	2–9
Dachshund, Wire	4.5	2–10
Dalmatian	5.8	1–9
Dandie Dinmont Terrier	5.3	1–7
Dobermann	7.6	1–13
Elkhound	6.0	1–13
Fox Terrier, Smooth	4.1	1–8
Fox Terrier, Wire	3.9	1–7
Foxhound	7.3	2–8
French Bulldog	5.8	1–10
German Shepherd Dog	8.0	2–15
Greyhound	6.8	1–13
Griffon Bruxellois	4.0	1–8
Hungarian Sheepdog	6.7	1–15
Irish Terrier	6.1	1–15
Kerry Blue Terrier	4.7	1–9
Lakeland Terrier	3.3	1–6
Manchester Terrier	4.7	1–7
Mastiff	7.7	2–15
Miniature Pinscher	3.4	1–5
Newfoundland	6.3	1–10
Norwich Terrier	2.8	1–6
Papillion	2.6	1–5

TABLE 3 – *continued*

Sizes of Litter for Various Dog Breeds

Breed	Average	Typical variation
Pekinese	3.4	1–6
Pointer	6.7	1–12
Pointer, German Short Hair	7.6	1–15
Pointer, German Wire Hair	8.1	1–16
Pomeranian	2.0	1–15
Poodle, Miniature	4.3	1–7
Poodle, Standard	6.4	1–10
Poodle, Toy	4.8	1–7
Retriever, Golden	8.1	4–14
Retriever, Labrador	7.8	2–14
Rottweiler	7.5	1–12
Samoyede	6.0	2–15
Schnauzer, Standard	5.1	1–12
Scottish Terrier	4.9	1–6
Setter, English	6.3	2–11
Setter, Gordon	7.5	1–14
Setter, Irish	7.2	1–15
Shetland Sheepdog	4.0	1–7
Shih Tzu	3.4	1–7
Siberian Husky	5.9	3–11
Spaniel, Cocker	4.8	1–10
Spaniel, English Springer	6.0	1–9
Spaniel, Irish Water	8.0	2–15
Spaniel, King Charles	3.0	1–5
St. Bernard	8.5	3–15
Welsh Corgi, Pembroke	5.5	3–8
Welsh Terrier	4.0	2–6
West Highland Terrier	3.7	1–8
Whippet	4.4	1–7

a lethal combination of genes, but many are due to uterine factors. If the ovary has released an unusually large number of eggs, the uterus could be overcrowded and some foetuses die, to ensure that the remainder have space to develop. Another common cause is when two eggs implant too closely together; one may weaken and die. The percentage is small and probably similar to that for still-births. In rare instances the proportion of still-births or foetal atrophy may show a sudden rise. If this happens genetic cause should be suspected.

The expected ratio of male to female pups is 1:1, but in reality marginally more males are born than females. In a survey for this book of data published in the canine literature and totalling approximately 6,695,000 pups of all breeds, the sex ratio was 103.4 males per 100 females or 50.8 per cent males. For practical purposes this may be regarded as a 1:1 ratio. A tendency for some breeds to have a significantly higher or lower sex ratio than the average can be noticed, but the reason, other than chance deviation bias in reporting, or registration of pups, is obscure. The tendency is small and would only be detected for large numbers of pups. The trend, such as it is, for larger breeds to have slightly higher sex ratios. No real association has been noted with age of sire or age of dam.

It is to be expected that longevity will vary between breeds as the cumulative outcome of several factors. On the one hand, there will be genetic endowment of a vigorous constitution and, on the other hand, the liability for various maladies. There will also be the chance exposure to infectious diseases. The average life of a dog appears to be in the region of 10–13 years, with a few individuals surviving to the exceptional ages of 15 or more years. Medium sized breeds live the longest, on the average, with the smaller breeds falling into second place. The largest breeds fare the worst, with average life-spans as low as 7–10 years. The genetic basis to longevity expresses itself as differences of life-span between breeds and even between strains within a breed. For fuller details see Comfort (1956, 1960).

3.

Development: Nature and Nurture

A DISTINCTION should be made between the genetic constitution and subsequent development of the individual. The former is fixed at fertilization and is the sum of the parental contributions to the individual. It is, of course, a truism that development has yet to occur, and this is where the effects of the environment enter the picture. It is worthwhile to consider what is meant by environment, for geneticists probably employ the term in a wider sense than the ordinary person.

The environment may be defined as all non-genetic influences to which the individual is exposed, particularly during the vulnerable periods of development and growth to maturity. There is the prenatal environment, as provided by the uterus. For example, the small size of the "runt" of the litter could be caused by a deleterious gene but also by an unfavourable position within the uterus which inhibited normal growth. There is the postnatal environment, as provided by the mother's care and nursing ability. A poor flow of milk can check the growth of puppies at a critical stage so that they fail to reach their optimum size.

However, the environmental factors which usually spring to mind are the diet, surroundings and temperature. The former is of first importance, of course, since growth is dependent upon a nourishing and adequate diet. Inferior diet or even an incorrect feeding schedule could have an adverse effect upon puppies and young dogs. Also, proper and sufficient exercise is important for the development of muscles and healthy bodies. Cold, damp and ill-ventilated conditions could interfere with growth and encourage the outbreak of disease; as would also overcrowding of young and half-grown dogs due to inadequate housing or thoughtlessness. Note that all of these effects are entirely detrimental or negative in effect. If a dog is to realize its full potential, conditions of feeding and husbandry must be at least adequate, but preferably of the highest order. Anything less could have a hampering effect of one sort or another.

At the other extreme, features such as coat colour and quality are only slightly affected environmentally. Black poodles may have an off-black coat if fed a diet consisting of plant protein but a jet-black coat if fed a diet rich in animal protein (Hollis and Whitney, 1957). It is known that many mammals may develop coats deficient in pigment granules if fed diets very low in protein. The poodle situation, however, may represent a special case. The hairs of the coat are not affected. Rich feeding may impart a bloom to the coat, but this is due mainly to the dog being in excellent health and condition. Grooming is important in this respect, so long as it is appreciated that this is only a cosmetic effect. Everyone likes to see a well-turned-out dog.

A rather subtle environmental effect is variation due to accidents and irregularities of development which cannot be easily ascribed to specific causes. In general, variation of this nature is small, but for some characteristics it may be substantial. For example, while most of the variation of intensity of yellow or red colour is genetic, a small part could be non-genetic, i.e. due to embryonic development of the individual dog. Of greater significance is the variation of amount of white or piebald pattern. While most of the variation is genetically controlled, a proportion is due to idiosyncrasies of development, so that no two white patterned dogs are ever identical. Some inbred strains may seem surprisingly uniform, and are recognizable because of their characteristic markings, but nevertheless it is extremely rare for two pied-marked dogs to be alike.

Thus, to round off this section, the individual dog owes his appearance to an interplay of heredity and environment. Some characters are only slightly modified by the environment, while others can be more sharply affected. Growth and development is the most obvious example of the latter. A superlative dog can be given that something extra by careful attention, feeding and grooming, but no one has yet been successful in changing a mediocre dog into a champion. However, it is easy to do the opposite, cripple the chances of a good dog by poor management.

The Germ Cell Lineage

The dog comes into the world as a union of two germ cells, the egg from the bitch and the sperm from the dog. This union, or fertilization, occurs within a thin tube of tissue, the oviduct, down which the eggs are impelled after release from the ovary. A successful copulation will have deposited vast numbers of sperm in the neck of the vagina, and these will make their way to meet the oncoming eggs. Despite their enormous numbers, few of the sperms will arrive in the vicinity of the eggs and only one will actually fertilize each egg.

The fertilized egg is a tiny speck of colourless protoplasm, but growth and development is rapid within the comfortably warm uterus. By about the 19th day, the eggs implant themselves to the wall of the uterus, spaced so as to avoid overcrowding. The continued growth transforms the egg into a foetus, scarcely recognizable as a pup at first but definitely shaping up. After an elapse of about 63 days, the pups are born.

Since the pup is created by the union of germ cells, it is obvious that their contents must be the material link between successive generations. The egg and sperm differ greatly in relative size, the former being spherical and full of nutriment to carry it over the few days from fertilization to implantation. The sperm is a microscopic tadpole-shaped cell, with a head and long tail to enable it to swim in the uterine fluids. Inside the egg and sperm is a nucleus, and the function of fertilization is to ensure that the two nuclei (one from the egg and one from the sperm) can come together and fuse. The important constituents of the nuclei are the chromosomes, the actual bearers of the genetic material. These are the bodies which give life to the new individuals.

As the foetus and the eventual young pup grows and develops, the myriads of cells involved are busily constructing the organs of the body and performing the many physiological functions necessary for a healthy existence. All of these are genetically controlled, or "programmed" to use a modern idiom. It is when something goes wrong in the programming that a genetic disease makes an appearance. These will be discusssed more fully in a subsequent chapter. Development proceeds by a number of semi-autonomous pathways, subject to the restriction that the overall pattern of growth must be harmonious. This is how the foetus develops into a graceful dog, with strong bones and rippling muscles.

A fundamental divergence of growth is between those cells which are destined to produce the body and those which are destined to form the

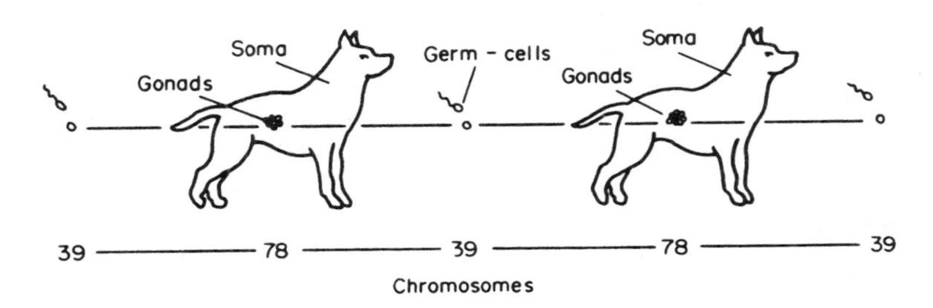

FIG. 1. A fundamental distinction is made between the germ-cell lineage and the body or soma. The germ cell (sperm and egg) is the bridge between the generations, transmitting the hereditary material (chromosomes) and enables the germ line to continue indefinitely. The germ cells each contain 39 chromosomes and, by fusion, restore the 78 chromosomes of the body cells.

germ cells. It is possible, therefore, to postulate a sequence of cells directly-concerned with reproduction, the "germ cell lineage". The reproductive organs of the male and female are the testes and ovaries, respectively. These are also known as the gonads, especially when each is in a rudimentary state before becoming differentiated into the male and female organs of the adult. The rudiments of the gonads are laid down early in development and are largely independent of the remainder of the body.

The concept of the germ cell lineage is important since it is fundamental in visualizing what is meant by inheritance and what is actually inherited. The germ cell lineage may be imagined as the passage of cells in an unbroken line as gonads–germ cells–gonads–germ cells, bridging generations as the egg and sperm. This should not be taken as implying that the genetic constitution of the gonads and germ cells is unchanging. This is far from the truth, for the genetic constitution is the chromosomes and the genes upon them, these being subject to mutation and recombination, as well as the forces of selection.

The concept does highlight the point that any mishap or deliberate mutilation of the body will not affect the germ cells and, therefore, will not be inherited. This exposes the fallacy of acquired inheritance which presupposes that an environmental influence affecting the body will also reach the germ cells and be inherited. This is not so, and the cruder forms of acquired inheritance have long been discredited. The docking of tails or cropping of ears *per se* will not produce pups without tails or ears. This is shown by the fact that the docking or cropping has to be performed each generation.

An understandable fear is where a disease has so affected a pregnant bitch that she produces diseased or defective offspring and that the condition will be inherited. Nevertheless, neither the germ cells of the mother nor of the offspring will be affected. If the disease can be irradicated, the bitch should produce normal pups in future litters and the offspring upon maturity will reproduce normally.

Rather akin to the above is the once commonly held belief that a valuable pedigree bitch would be spoilt for future breeding if she was accidentally mated by a mongrel dog. The belief is that her pups will be mongrels however she is subsequently mated. This is nonsense, as a little reflection will reveal. The male sperm is very short-lived and every mating is quite independent of others. The reverse possibility is also false. Availing oneself of a superior male for a first mating will not guarantee the production of above average pups in subsequent litters by inferior studs. The use of one male is incapable of influencing the progeny of another. This can be shown in cases of accidental double matings by two males. Usually all of the pups will be sired by one male, but occasionally mixed litters occur when the pups can be seen to be sired by one male or the other, and not pups showing a mixture of traits from both sires.

Cell Division

The chromosomes are so fundamental that consideration should be given to how they behave in the nucleus of cells. Since they are the bearers of the genes, they will in fact dictate how the genes are inherited. The chromosomes are the link between generations, contained within the nuclei of the fusing germ cells. Thus, a working knowledge of the behaviour of the chromosomes in the formation of the germ cells is useful for an understanding of heredity.

The mechanism by which the chromosomes multiply and form new cells is not particularly complicated. In its simplest form, a cell may be imagined as a small sac of colourless watery fluid (protoplasm) surrounded by a thin membrane (cell wall). Inside the cell is the nucleus, the contents of which are enclosed by another membrane (nucleus wall). One of the functions of the nucleus is to house the chromosomes instead of having these dispersed at random thoughout the cell.

Ordinary cell division, which is responsible for growth and formation of all the organs of the body, is known as mitosis. Mitosis is the mechanism by which cells multiply and build up tissue. When the cell is ready to divide, the wall of the nucleus dissolves away, releasing the chromosome into the cell. These are initially jumbled up, but quickly align themselves across the centre of the cell. Presently they split into two. This is the simplest way of visualizing the process, for the chromosomes do not actually split; what happens is that each one manufactures a replica of itself. This latter peels off, giving the impression of splitting. The old and new chromosomes travel to opposite ends of the cell as a group, whence a membrane forms around each group, reconstituting two nuclei. The division is completed by the formation of a thin wall across the middle of the cell, making two new cells identical to the original. The cells of the dog contain 78 chromosomes and each one behaves as above, independently yet in perfect unison, to form daughter cells. The process is repeated over and over again to build up tissue which evenutally becomes organs.

The mitotic divisions are responsible for the formation of the gonads, these being built up at the appropriate stage of embryonic development. The fact that they are founded so early in life does not mean that they are functional; this does not occur until puberty is reached many months after birth. At this time under the influence of hormones, the gonads become active and commence to produce germ cells. This is accomplished by further cell divisions, similar but of a different nature to mitosis. This modified form of mitosis is known as meiosis and is an essential prelude to the formation of germ cells.

As in mitosis, meiosis begins by dissolution of the nuclear membrane and the orderly line-up of the chromosome across the centre of the cell.

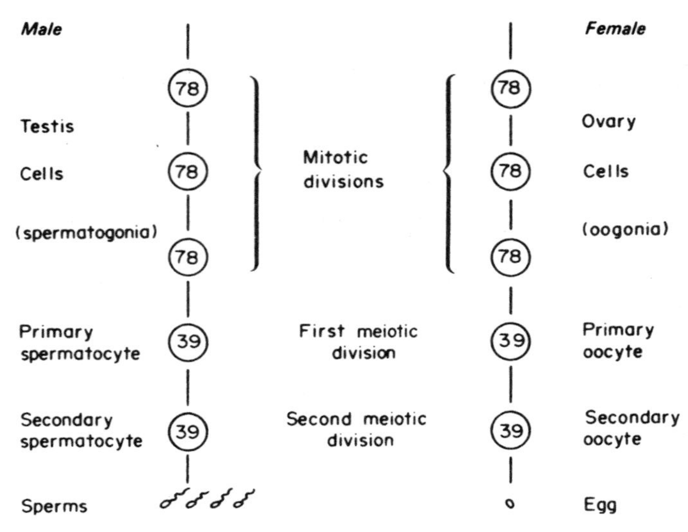

FIG. 2. Reduction of chromosome number occurs in the gonads or reproductive organs of the male (dog) and female (bitch). Although the processes are given different names in the sexes, these are fundamentally identical. The chromosome number is halved at the meiotic divisions.

However, instead of behaving independently, they come together in pairs. The pairs of chromosomes do not join up, although the pairing is very intimate. Careful examination has revealed that the pairing is not at random but involves those chromosomes of similar size and shape. From this fact, it is clear that the 78 chromosomes are in reality 39 pairs; that is, two of each kind of chromosome. The pairing is only momentary, for the chromosomes soon commence to move away from each other to opposite ends of the cell. Grouping together, they are soon enveloped by a nuclear membrane, and the whole process is completed by the throwing up of a cell wall to produce two new cells. Note the close parallelism with mitosis, the difference being that the chromosome do not initially split but come together in pairs. The outcome is that the daughter cells contains 39 chromosome instead of the usual 78. A second meiotic division now occurs, which is essentially a mitosis, the chromosomes splitting in the usual manner and the two daughter cells each receiving a complement of 39 chromosomes.

The important meiotic division is the first, and the significant aspect is the efficient reduction in the number of chromosomes. The full complement of 78 chromosomes is known as the diploid number, but this consists of double the basic, or haploid, number of 39. The daughter cells produced by meiosis are distended to become germ cells, each containing 39 chromosomes as opposed to 78 for the body cells. The transformation is more or less direct in the male (spermatogenesis), yielding sperms, but is less direct for the female (oogenesis), although the end result is the same, namely eggs. The union of two germ cells, each with 39 chromosomes, restores the diploid number of 78 found in the body cells.

The necessity of halving the number of chromosomes in the germ cells should now be apparent. If this did not occur, the number of chromosomes would double for each generation, which is absurd. Moreover, the substitution of meiotic divisions prior to the actual production of the germ cells is an elegant mechanism for maintaining the constant number of chromosomes. Thus, there is an alternation of 39-78-39-78-39 for germ cells and body tissues. The arrangement of the chromosomes in pairs is a vital part of the process. It is not sufficient that the germ cells should receive any random collection of 39 chromosomes, but the same basic 39 in each cell. The alignment of paired chromosomes ensures that one of each pair is despatched to opposite ends of the cell. Although the chromosomes may be jumbled together when the nuclear membrane dissolves away, the meiotic division effectively sorts out the pairs. The haploid–diploid alternation of number is confined to the germ-cell lineage, the body cells invariably containing 78 chromosomes, except for accidents of division. These latter do occur on rare occasions with odd results, but these need not trouble us at this stage.

Determination of Sex

There is an exception to the statement that the chromosomes of mammals occur as matched pairs. Two of the chromosomes, although they behave as a pair during meiosis, differ in size. For some species the difference is not obvious, but in the majority the difference is marked. This is the situation in the dog, for one of the pair of chromosomes is very large while the other is the smallest chromosome of the set. These are the sex chromosomes, so-called because the female always has two large chromosomes while the male always has one large and one small chromosome. The two chromosomes are known as the *X* and *Y*, for the large and small, respectively.

Confirmation that the *X* and *Y* chromosomes are a pair is revealed by their behaviour at meiosis in the male. It was observed that the *X* and *Y* invariably ended up at different poles of the cell. Thus, in the male with the *XY* constitution, half of the spermocytes have *X* and half the *Y* chromosome. This means that the sperms which develop from these cells will not be completely identical. In other words, two sorts of sperm will be produced in equal numbers.

The situation is different for the female. These have two *X* chromosomes which pair at meiosis in the usual manner and are consigned to opposite ends of the cell. All of the eggs which result will contain an *X*. It follows, consequently, that at fertilization the constitution of the individual as regards the *X* or *Y* will depend upon the sperm. The egg will contribute an *X* while the sperm will contribute an *X* or a *Y*, to give individuals which are either *XX* or *XY* respectively. These will be produced in equal numbers or, as sex statistics are usually portrayed as a 100:100 ratio of male:female.

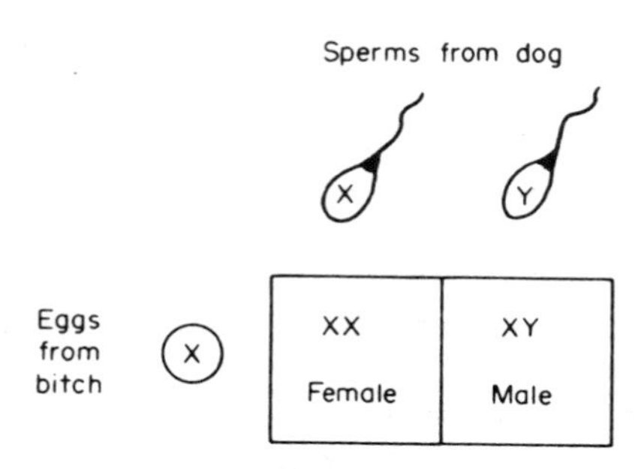

FIG. 3. In the normally developed pup, sex is determined by the *X* or *Y* chromosomes present in the sperms. The sperms, being either *X* or *Y*, fuse at random with the eggs, all of which are *X*, to produce near equality of the sexes.

Therefore, it is possible to say that sex is determined by the male by virtue of the fact that he is producing two sorts of sperm. Since the *Y* is found only in the male, it is often referred to as the male chromosome. Conversely, the *X* may be referred to as the female chromosome. Although it is generally true to say that sex is determined at fertilization, it is more correct to speak of the sex potential being determined at this time. The fertilized egg has yet to develop into a normal female or male.

Fortunately, accidents of sex development occur only rarely, usually of such a nature that the individual may possess sex organs partly of one sex and partly of the other. These unfortunate individuals are known as intersexes, the implication being that they are neither of one sex nor the other. The occurrence of such animals are of concern to breeders and vets, and will be discussed later. An even more rare sex anomaly is where an animal possesses both male and female sex organs. These are known as hermaphrodites and are distinct from intersexes. Quite often the sex organs of hermaphrodites are not perfect, in which event it is possible for these to be confused with intersexes. It is also possible for intersexes to closely resemble hermaphrodites, when they are called pseudohermaphrodites.

The Chromosomes

The chromosomes occupy such a vital position in genetics and animal breeding that it is worthwhile to consider them more deeply. The dog possesses 78 chromosomes, which is actually 39 pairs. This is, as a point of interest, a large number for a mammal. The total number of chromosomes, together with a description of the size and shape of each one is known as the karyotype. The chromosomes are unremarkable in the dog, all of the ordinary chromosomes appearing as short rods. One pair is noticeably bigger than the others, but the remainder vary gradually in size from the largest to the smallest. Chromosomes of this form are said to be one-armed, rod or I-shaped.

Individual chromosomes represent blocks of genes, and a large number of chromosomes (as opposed to a few) implies a greater amount of recombination of genetic material. That is, a strong tendency to recombine, or shuffle at random, hereditary characters. This may be one reason why the dog shows such variation of conformation or, to put the matter another way, why breeders have been able to evolve so many different breeds. Having a large number of chromosomes cannot create innate genetic variation, but it can encourage the full exploitation of such variation as exists. To consider a wider aspect, a large number of chromosomes could have facilitated the wolf-jackal complex to successfully colonize so many habitats of the world. For a single complex of species to range from the subarctic to the tropics is no mean achievement.

If the ordinary chromosomes appear unremarkable, this is not the case for the sex chromosomes. These are exceptional in that they differ considerably in size. The *X* is the largest chromosome of the set, with two distinctive arms. In other words, the *X* often appears L- or V-shaped. On the other hand, the *Y* is a small chromosome, actually the smallest of the set. The large size of the *X* implies that it could carry many genes and, indeed, several sex-linked genes are known. Conversely, the small size of the *Y* implies that it has a limited function, evidently to initiate development of the embryo as a male.

The heredity of chromosomes is normally an orderly process, but mishaps do occur in a small number of instances. An interesting event is where one chromosome becomes attached to another. Since all of the ordinary chromosomes of the dog have one arm, the above event is immediately obvious as a chromosome which is larger than normal and which will have two arms. Another clue will be the reduction of the number of chromosomes from 78 to 77. Dogs with such a chromosome are often of normal appearance because there is no gain or loss of genetic material. The usual effect of chromosome fusion is a reduction of fertility. Several cases have been reported among different breeds of dog. Frequently, the attachment is between the medium and smaller chromosomes, if only because these are the most numerous.

No less than eight instances of chromosome fusion have been reported for the dog. It has not been possible to identify the chromosomes involved for six of the fusions because these were observed before the techniques

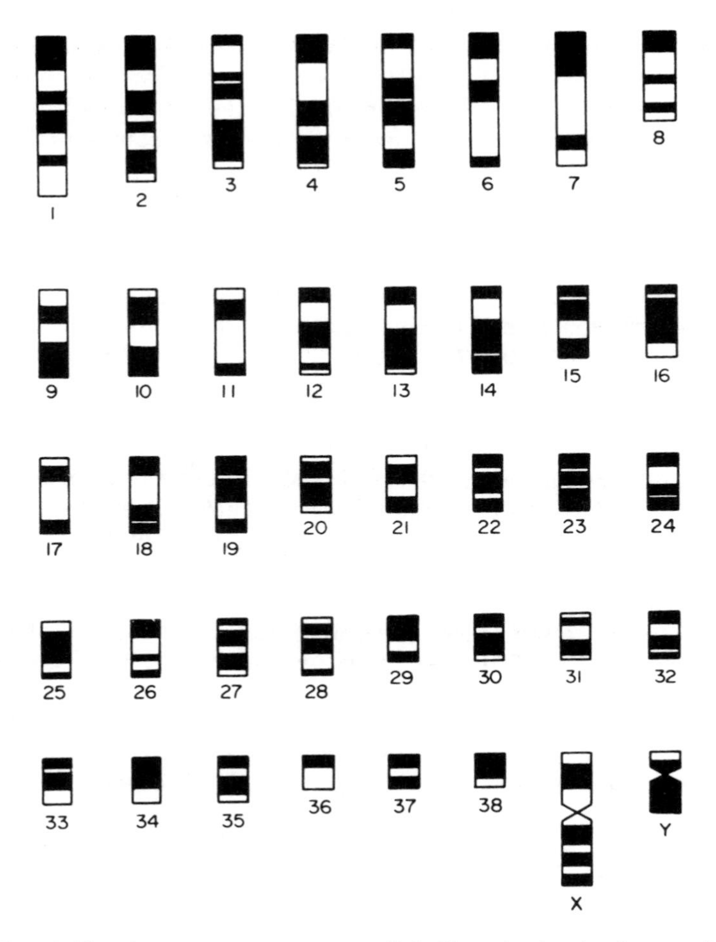

FIG. 4. The chromosomes occur as small darkly stained rods of matter in the nucleus of cells. Normally, most of the chromosomes appear as featureless bodies. However, by special techniques, it is possible to reveal the chromosomes as short banded rods. The individual chromosomes can be identified by the number, relative width and intensity of the bands. In the above example, only the number and widths of the bands are shown. It should be noted that no two chromosomes are exactly alike.

of chromosome banding had been perfected (see Fig. 4). The first fusion for which the chromosomes could be identified involved the medium sized number 13 and the smaller number 31 (Larsen *et al.* 1979). The fused chromosomes were inherited as a unit among the cross-bred offspring of a Golden Labrador bitch. The carriers of the fusion showed no signs of abnormality and the fertility of the offspring did not seem to be affected.

The second fusion where identification was possible involved the large chromosome 1 and the small chromosome 31 (Mayr *et al.,* 1986). The heredity of the fusion was not observed directly but 7 out of 25 individuals of a strain of Poodles possessed it. This implied that the fusion was being genetically transmitted. All of the carriers were healthy and quite possibly there was little or no effect on the fertility.

The consequences are more serious when the chromosome mishaps involve the sex chromosomes. The normal development of the sex organs are disrupted and one of several kinds of sex anomaly is the result. The type of anomaly may not be apparent until a post-mortem investigation is performed, but some idea is often possible from a careful clinical inspection. Any dog which is sterile, has a high proportion of still-born pups or shows abnormalities of the external sex organs should be taken for examination by a veterinary surgeon.

Animals with various sex anomalies are known as intersexes or pseudohermaphrodites. They usually show a mixture of male and female internal sex organs, but sometimes predominately one or the other. True hermaphroditism, where the sex organs of both sexes are present, is very rare. Many of the intersexes arise from accidents of development and no obvious cause can be found. In certain cases errors in the transmission of the sex chromosomes can lead directly to a sexual anomaly. As examination of the chromosomes is becoming more a matter of routine in veterinary practice, several categories of anomaly are being defined.

A common chromosomal intersex is where the individual is a mixture of *XY* and *XX* cells (a *XY/XX* mosaic as these are termed). Some cells of the body will be *XY* (male) and others *XX* (female). It is easy to imagine that these curious animals could possess sex organs of both sexes, albeit poorly developed, so that they are neither one sex nor the other. Other mosaics can be *XX/XXY,* that is a proportion of the body cells contain an extra chromosome, the *Y*. Such a mixture is usually sufficient to prevent normal embryonic sex development even when the proportion of *XXY* cells was small (Dain and Walker, 1979).

A particularly interesting case is the *XXY* male dog. Here, the animal contains an extra chromosome, a *Y*, giving a total complement of 79 in all cells. Externally, the dog is a male but sterile due to an absence of sperms in the semen (Clough *et al.*, 1970). These individuals show the strong male inducing effect of the *Y*, in spite of the presence of two *X* chromosomes. On the other hand, the lack of sperm reveals that normal sexual development has been disrupted. The testes failed to develop completely and are flaccid upon palpation.

Another interesting sex chromosome anomaly is the *XO* female which has only one *X* chromosome. The individual is externally normal except that it may be somewhat undersized and sterile. No cases have yet been reported for the dog but several have been found in the cat and in a number of other animals. It may be worthwhile to have the chromosome complement examined for any bitch which fails to show an oestrous cycle.

Sex development, in fact, is more complicated than simple fertilization of an egg by either an *X* or *Y* sperm. These give the potential sex, so to speak,

TABLE 4

The Number of Chromosomes for Members of the Dog Family, the Canidae

Common name	Zoological name	Chromosome number
Arctic fox	*Alopus lagopus*	48
Bush dog	*Speothus venaticus*	74
Cape hunting dog	*Lycaon pictus*	78
Coyote	*Canis latrans*	78
Crab-eating dog	*Cerdocyon thous*	74
Dingo	*Canis dingo*	78
Domestic dog	*Canis familiaris*	78
Fennec fox	*Fennecus zerda*	64
Grey fox	*Urocyon cinereoargenteus*	66
Hoary fox	*Dusicyon vetulus*	74
India fox	*Vulpes bengalensis*	60
Jackal	*Canis aureus*	78
Japanese raccoon dog	*Nyctereutes viverrinus*	42
Maned wolf	*Chrysocyon brachyuras*	76
Raccoon dog	*Nyctereutes procyonoides*	42
Red fox	*Vulpes vulpes*	34 (38)
Red wolf	*Canis rufus*	78
Ruppell's fox	*Vulpes ruppeli*	40
Sand fox	*Vulpes pallida*	34 (38)
Short-eared dog	*Atelocynus microtus*	76
Wolf	*Canis Lupus*	78

rather than a guarantee of normal development. The above cases should bear this out. Just how complicated the processes can be is outlined by Selden *et al.* (1978) and Meyer-Wallen and Patterson (1988) with reference to the dog, and people who wish to follow up this topic should read this article. Abnormalities of sex development are of concern to breeders, and the more cases which are referred to veterinary clinics, the greater will be the understanding of the underlying causes.

Inspection of the chromosome numbers for the various species listed in Table 4 will reveal that the coyote, dingo, dog, jackal and wolf possess the same number. The similarity is indeed even more remarkable when it is realized that the sizes and types of chromosome which constitute the karyotype are identical in all species. This is the fundamental reason why hybrids can be obtained. It indicates, moreover, that the various species have not diverged genetically as far as one might anticipate by appearance alone. There is no reason why the species should not be regarded as a species complex: a group of species on the road to full speciation. This view would be compatible with the usual zoological nomenclature for the species, but emphasizing the evolutionary implications.

By the same standard, the great difference in number of chromosomes for the red fox, compared with the dog or wolf, indicates that these are distinct species, in spite of their apparent physical resemblance. the difference of chromosome number is one reason for doubting the authenticity of alleged hybrids between the dog and fox. The similar appearance could be an example of convergence evolution, where two distinct species come to resemble each other because they share a common habitat and are subject to similar selection pressures.

4.

Elements of Heredity

It is quite feasible to breed dogs without any deep understanding of the laws of heredity. In fact, many breeders do in fact breed dogs in this manner; quite successfully in many cases! The reason may be made clearer by an analogy. A person may be a proficient driver without being unduly concerned with what is happening under the bonnet. However, if a person has taken the trouble to learn what is happening, he or she may not be a better driver as a consequence, but certainly a more knowledgeable one. Since knowledge usually leads to greater understanding, it is likely that in many instances the person will actually be a better driver!

Mendelism and Genetics

The basic laws of heredity were discovered (or, perhaps, more accurately, deduced) by Gregor Mendel as long ago as 1866. Alas for such inspired reasoning, the biologists of his age were not ready to accept Mendel's ideas. The apparent simplicity of Mendel's logic probably worked against him. People thought: surely heredity is not as simple as this? However, by 1900 fresh minds had grappled with the problem and Mendel's work was acclaimed by three biologists, independently, and more or less simultaneously.

It is well known that Mendel carried out his work with the vegetable pea, but once the principle had been grasped, people were eager to discover if his ideas applied to other plants and, particularly, to animals. It was soon found out that they did, and with remarkable similarity. This general applicability meant that Mendel had uncovered a fundamental scientific law. In 1906 William Bateson proposed that the study of the laws of heredity (hitherto referred to as Mendelism) should be known as genetics and a fledgeling new science was born. From the humble beginnings of a man pottering around his vegetable garden, genetics has grown to such an extent that no branch of medicine, biology, plant or animal breeding can now escape its impact.

There is no grand design in science. This implies that the application of new ideas can be very uneven. In some quarters, genetics has received more than its fair share of opposition from established practice. Certain aspects of animal breeding is a case in point. Animal breeding has an ancient history and has a rich quota of legend based on all sorts of queer prejudices. This delayed the application of genetics to some extent, especially those farthest removed from economic pressures.

The art of successful breeding is selection; careful choice of breeding stock to beget superior offspring in future generations. This process had been practised long before Mendel's ideas began to take root. In fact, traditional animal breeding was carried on for several decades before genetics caught up with it. Animal improvement required a comprehensive theory of the effects of selection, inbreeding and cross-breeding if it was to gain general acceptance. Ideas were prevalent enough, but many were little more than firmly held opinions and had little rational basis. Genetics has changed all this, for one can now ask what if I do such and such, should I expect this and that or perhaps not? Genetics can attempt to answer many difficult questions or, if this is impossible, it can at least offer a rational basis from which the right questions may be asked.

Mendal termed the unit of inheritance a "factor". For a long time the units were referred to as "Mendelian factors" or "unit characters". Now all of these terms have been dropped in favour of "gene". The term gene is as important in genetics and animal breeding as the term atom is to chemistry and physics. Individual atoms may be invisible, but look at what can be accomplished with them! So be it with genes, but with a difference. Individual genes may be too small to be seen, but *en masse*, neatly packeted as part of the chromosome, they can be seen through the lens of a powerful microscope. Suitably dyed, the chromosomes appear as rods or dots of material in the nuclei of cells. It is immensely satisfying to be able to point to a material object and say that these are the determinants of heredity.

Laws of Heredity

The principles of inheritance are relatively simple and may be summed up by two basic laws: (1) the law of independent separation of the genes in the germ cells and (2) the law of independent assortment of characters in the individual. These bald statements will be amplified with particular reference to their application to dog breeding.

It may be wondered if the very simplicity of the basic laws may operate to their disadvantage. This may be true to some extent, but one should not be taken in by the apparent simplicity. A step-by-step approach from humble beginnings is useful for learning most subjects and has certainly proved itself for genetics. In considering a complex animal like the dog, the procedure would be to seek out elementary attributes for explanatory purposes. In genetics these are known as unit characters or, more simply, as characters. It may be felt that such an approach would be arbitrary, but this need not necessarily be so. As the subject unfolds, additional attributes can be considered until the overall picture is as complex as one may wish. In this manner the bare bones of heredity can be clothed, so to speak.

Genes and Alleles

The genes are the units of heredity and are arranged on the chromosomes like beads on a string. This is a time-honoured means of visualizing the genes and is still useful as a mental image for relating the gene to the chromosome. In essence, the comings and goings of the genes mirror the behaviour of the chromosomes at meiosis. At this junction, the behaviour of genes is of greater concern than that of the chromosomes, and little more will be said about them; but it is wise to remember that behind the genes lie the chromosomes.

The chromosomes, hence the genes, multiply during meiosis by a process of self-copying which is remarkably exact. The copying is so excellent that many thousands of daughter chromosomes are produced without a mistake occurring. However, on rare occasions an inexact copy occurs at the level of the gene. The event is known as "mutation" and the new gene as a "mutant". A mutation means that a gene has come into being with a changed physiological function. It still occupies the same position in the chromosomes as the original gene and will be inherited as an alternative in heredity. For this reason the original and the new gene are spoken of as "alleles", meaning "other". It is the property of pairs of genes being inherited as alternatives to each other which makes sense of much of elementary genetics.

The occurrence of mutation is so infrequent that, for most practical purposes, the event can be ignored. It is an indication of the rarity of mutation that the number of mutants known for the dog is so few, in spite of the millions of animals bred all over the world. Only about two dozen or so mutants for coat colour or quality have been reported. There are several reasons for this, of course, aside from the genetic rarity. New mutant colours may not be recognized as such or, if they are, they may not be attractive enough for them to be preserved.

The coat colours of the dog may be employed as examples of the principles of heredity. It should not be imagined, however, that genetics is concerned solely with such characters. Far from it as a point of fact. The growth and maturity of the individual, its behaviour, reproductive capacity, liability to various diseases and afflictions are all governed in part by heredity.

Simple Inheritance

The simplest form of heredity is that of the transmission of a pair of contrasting characters. Black as opposed to chocolate or liver pigmentation is a good example of characters which are inherited as alternatives. That is, the coat is either black or chocolate; never black/chocolate mosaic or striped. Even a black coat faded by the sun to a brownish black is not likely to be mistaken for a chocolate.

The sort of questions which should be asked are: What would happen if black is mated to chocolate? What would be the colour of the first generation? What sort of colour or colours would be expected in the next generation? All seemingly innocent questions, yet they hold the key to understanding how heredity works. All of the answers are straightforward. The pups from a mating of black to chocolate will be expected to be black and if these are subsequently mated together a ratio of 3 black to 1 chocolate will occur in the second generation.

The follow-up question must surely be: How can these results be explained? This is where a return must be made to the concept of genes. Black pigmentation is the original pigment of the dog and chocolate arose from it by a mutation. That is to say, a great many years ago the gene for black pigmentation failed to reproduce itself exactly and gave forth a mutant allele for chocolate pigmentation. Given the opportunity, it will replace the black gene in the genetic constitution of the individual. It is convenient to have symbols to represent the different alleles in order to facilitate the discussion.

The symbol adopted for black pigmentation is *B* and that for chocolate is *b*. It is mandatory to represent genes and their mutant alleles by the same letter; one by the capital and the other by the small letter. The reason for this will be explained later. Since the chromosomes are present twice in the body cells and once in the germ cells, it follows that the genes are present twice in the individual (one gene on each chromosome) but singly in the germ cells.

Therefore, in terms of symbols, the black individual will have the constitution *BB* and be producing germ cells *B*. Similarly, the chocolate will be *bb* and producing germ cells *b*. Mating black to black or chocolate to chocolate will produce offspring *BB* or *bb*, respectively. As pure breeding dogs, this is to be expected. The point is that because the parents have two similar genes (*BB* or *bb*), the germ cells will be of one sort entirely (*B* or *b*, respectively).

If, now, black (*BB*) is mated to chocolate (*bb*), the germ cells from each will be *BB* and *b*, respectively. The first-cross (F_1, as it is termed) will be *Bb*. Experience with these crosses has shown the F_1 are black, hence gene *B* is said to be "dominant" to *b*; or that gene *b* is "recessive" to *B*. In other words, in the combination *Bb*, the effect of *B* will take precedence over *b*.

Dominance and recessive are two very frequently used words in genetics and it is advisable to memorize their meaning.

What happens when the F_1 are mated together to produce the next generation, F_2 as it is termed, is interesting. The F_1 has the constitution *Bb*; that is, two unlike genes, hence the germ cells will be of two sorts, *B* and *b*, in equal numbers. The germ cells from each F_1 parent will unite at random. The chances of fusion of a *B* egg with a *B* sperm (to produce offspring *BB*) is as likely as fusion of a *B* egg with a *b* sperm (to produce offspring *Bb*). Likewise, the chances of a *b* egg fusing with a *B* sperm (to produce offspring *Bb*) is as likely as a *b* egg fusing with a *b* sperm (to produce offspring *bb*). Summing the four types of offspring gives 1 *BB*, 2 *Bb* and 1 *bb*. Since *BB* and *Bb* are both blacks, the expected ratio would be 3 black: 1 chocolate. Thus the 3:1 ratio arises from the random combination of genes in the germ cells. These results can be derived by the aid of a diagram as shown by Fig. 5.

It may be wondered if this expectation has been experimentally verified and what were the results? Some breeding data are indeed available in the scientific literature. These showed that the F_1 were black, as expected, and the F_2 assorted into 285 black and 85 chocolate pups. The expected number of black and chocolate are 277 and 93, respectively, for a sample of 370. Exact agreement is not to be expected, of course, but the expected and observed numbers of black and chocolate puppies agree quite closely.

Germ-cells from black heterozygote *Bb* (rows) / Germ-cells from black heterozygote *Bb* (columns)	*B*	*b*
B	*BB* Black	*Bb* Black
b	*Bb* Black	*bb* Chocolate

FIG. 5. The expectation from mating two blacks heterozygous for chocolate; illustrating the basic 3:1 ratio for an F_2 generation. Note that among the black progeny, one-third will be homozygous *BB* and two-thirds *Bb*.

At this stage it is advisable to define several terms without which it is difficult to discuss genetics with precision and depth. The term germ cell may be continued to be used although the correct word is "gamete". The individual formed from the union of two germ cells is a "zygote". There are two sorts of zygotes: "homozygote", when the two genes of the individual are identical (*BB* or *bb*), and "heterozygote", when the two genes are not identical (*Bb*). Thus, it is possible to have a "homozygous dominant" (*BB*) and a "homozygous recessive" (*bb*). The phenomenon of dominance is shown by the choice of symbols; dominant genes are represented by capitals and recessive genes by small letters. By adhering to this rule, it is easy to remember which is the dominant and which is the recessive of a pair of genes.

The phenomenon of dominance heralds an important distinction. The two blacks *BB* and *Bb* are of identical appearance but of different genetic constitution. The outward appearance is referred to as the "phenotype" and the genetic constitution as the "genotype". Even for one gene pair, it will be appreciated how important is the distinction. For, although the homozygous and heterozygous blacks have an identical phenotype, they will breed very differently. The conclusion to be drawn from this is that the genotype of a given individual may not always be that implied by the phenotype.

The backcross of the F_1 to the chocolate is an instructive mating. The chocolate parent can only produce one sort of germ cell (*b*), whereas the F_1 will produce two (*B* and *b*). The random union of germ cells will produce *Bb* and *bb* progeny in the ratio 1:1 as shown by Fig. 6. This result indicates that the F_1 is indeed producing *B* and *b* germ cells in equal numbers. Also, the genotypes of the progeny are precisely known. The blacks must be heterozygotes *Bb*, while the chocolates will be homozygous recessives *bb*.

Again it may be wondered if experimental data on the assortment of the *B* and *b* genes may be available. This is indeed so, and the observed results published in the scientific press total 170 black and 170 chocolate pups. The expected numbers for a total of 340 pups is 170 black and 170 chocolate. Agreement between the expected and observed numbers is excellent

Germ-cells from chocolate *bb*	Germ-cells from black heterozygote *Bb*: *B*	*b*
b	*Bb* Black	*bb* Chocolate

FIG. 6. The expectation from mating the black heterozygote to chocolate; illustrating the 1:1 ratio for the backcross generation.

and substantiates the expectations. In parenthesis, it may be mentioned that the chocolate animals will breed true, even if derived from either the F_2 or backcross generations. The same sources which gave illustrative data on the two crosses also revealed that chocolate with chocolate matings produced 464 pups, all of which were chocolate. These are the sort of breeding results which convinced geneticists of the validity of Mendelian heredity.

The presence of both homozygous and heterozygous blacks in the F_2 means that is it impossible to write down the genotype by examination of the phenotype. The most that can be inferred is that at least one *B* is present. It is customary to denote ambiguities of genotypes of this nature by inserting a dash sign (–) in the genotypes. Writing *B–* is to indicate that the actual genotype may be either *BB* or *Bb*. The genotype of the homozygous recessive can always be written down in full, since the genotype cannot be other than *bb*. Similarly, the blacks in the backcross of the F_1 to the chocolate cannot be other than *Bb*.

The inheritance of black versus chocolate is straightforward. The *b* allele disappeared in the F_1 but reappeared in the F_2 and backcross in its original form. Neither the *B* nor the *b* gene contaminate each other as a consequence of being in close proximity in the F_1 individual. Neither the black nor chocolate phenotypes will display a blend of characters. The example has illustrated the law of independent separation of the genes in the germ cells, the first of the two laws given in an earlier section.

Checkerboard Diagrams

One of the problems of animal breeding is that one has usually to deal with more than one character at a time. It would be useful, therefore, to gain insight into this situation. This may be achieved by considering the simultaneous inheritance of two pairs of genes.

The pigment granules of the hair are regularly and closely packed so that a dense colour is produced to the human eye. Black, for example, is obviously a dense colour. However, a mutant allele has occurred which dilutes the colour so that a blue-grey is produced. The effect is created by the pigment granules being less regularly packed, with clear spaces, thus "diluting" the colour, at least to the eye. The normal gene is called dense (*D*) and the mutant allele blue dilution or dilution (*d*) for short. The dilute allele is inherited as a recessive to dense (remember the rule for dominance in assigning symbols). The inheritance of *d* would be identical to that of *b*, as may be verified by substituting *D* for *B* and *d* for *b* in Figs. 5 and 6 and deriving the expectations. The *DD* and *Dd* individuals are black in colour, while *dd* is blue.

The *D* and *d* genes are distinct from *B* and *b*; hence, it is possible to have the genes in combination. This means that the effect of each gene must be taken into account in the same animal. For instance, black is a dense colour, hence the genotype is *BBDD*. Chocolate is also a dense colour; therefore, the genotype is *bbDD*. Blue is a dilute colour, with the genotype *BBdd*. But what colour is *bbdd*; namely, chocolate dilute? Not to maintain the suspense, the colour is lilac-grey as found in the Weimaraner breed.

What are the expectations when two pairs of genes are involved in the same mating? This may be exemplified by the cross of a chocolate (*bbDD*) with a blue (*BBdd*). The germ cells will contain one of each pair of genes, *bD* from the chocolate parent and *Bd* from the blue. The union of these germ cells gives an F_1 of *BbDd*; namely, the restoration of four genes as in the parents but of a totally different combination. In appearance the F_1 will be black, because *B* and *D* are dominant to *b* and *d*, respectively.

The only problem in finding the F_2 from this mating is to derive correctly the constitution of the germ cells of the F_1. The guiding principle is that the germ cells must contain one of each pair of genes, but only one of each pair. The genes must also separate at random. This means the *B* is as likely to combine with *D* or with *d*, to give germ cells *BD* and *Bd*. Similarly, *b* is as likely to combine with *D* as with *d* to give germ cells *bD* and *bd*. In total, four different germ cells are produced, *BD*, *Bd*, *bD* and *bd*. The F_2 will consist of individuals from the random union of these germ cells. The expectations could be derived by describing how the germ cells are likely to meet with each other, as in the case for one pair of genes. However, this is a laborious method and, fortunately, the expectations can be quickly and more accurately found by the aid of the diagram of Fig. 7.

The diagram of Fig. 7 is known as a checkerboard, and is a valuable device for working out the expectations. The diagrams of Figs. 5 and 6 are checkerboards, but these are so simple that they fail to demonstrate the usefulness of the method. Checkerboards are easy to construct and faithfully reproduce the random assortment of the genes and the complete array of genotypes. Given the latter, the phenotypes follow automatically by considering the dominance relationships of the pairs of genes.

The checkerboard is constructed according to the following rules. All of the possible different germ cells are written along two sides of a large square; those from one parent along the top and those from the other down the left side. Horizontal and vertical lines are drawn to divide the square into as many rows and columns as there are different germ cells. Within each smaller square are written the gene symbols (representing the germ cells), at the side of the row and head of the column in which the square resides. To complete the diagram, it is only necessary to write in the phenotype. Summing these will result in the expected ratios. Checkerboards provide valuable exercise in the manipulation of genes and there should be no hesitation to make full use of them.

In the present example, the assortment of the genes produces the following phenotypes: 9 black (*B–D–*), 3 chocolate (*bbD–*), 3 blue (*B–dd*) and 1 lilac (*bbdd*). These are the ratios to be expected if the genes are combining at random. Note the dashes in the above genotypes to indicate that most phenotypes can have both homozygous and heterozygous forms. In fact, 4

Germ - cells from heterozygote *BbDd* (columns); Germ - cells from heterozygote *BbDd* (rows)

	BD	*Bd*	*bD*	*bd*
BD	*BBDD* Black	*BBDd* Black	*BbDD* Black	*BbDd* Black
Bd	*BBDd* Black	*BBdd* Blue	*BbDd* Black	*Bbdd* Blue
bD	*BbDD* Black	*BbDd* Black	*bbDD* Chocolate	*bbDd* Chocolate
bd	*BbDd* Black	*Bbdd* Blue	*bbDd* Chocolate	*bbdd* Lilac

FIG. 7. The checkerboard diagram for working out the expectation for assorting genes. The expectation from mating two doubly heterozygous black, illustrating the 9:3:3:1 ratio for two pairs of genes. Note the various homozygotes and heterozygotes among the different colours.

of the 9 blacks can be heterozygous for both *b* and *d*, and 4 others heterozygous for either *b* or *d*. It follows, however, that any individual displaying a recessive phenotype must be homozygous for the relevant allele. Thus, the lilac-grey (*bbdd*) must be a double homozygote. Similarly, the chocolate will be homozygous for *b* and the blues will be homozygous for *d*.

An interesting mating would be that of the F_1 *BbDd* with the double recessive *bbdd*. The expectations are shown by Fig. 8. The same four colours are produced, but now in the ratios of 1:1:1:1. The perspicacious reader will observe that these ratios are a direct extension of the 1:1 ratio of Fig. 6. The four phenotypes reflect the random separation of the genes in the F_1 individual. The lilac-grey parent will be producing only one sort of germ

Germ - cells from heterozygote *BbDd* (columns); Germ - cells from lilac *bbdd* (row)

	BD	*Bd*	*bD*	*bd*
bd	*BbDd* Black	*Bbdd* Blue	*bbDd* Chocolate	*bbdd* Lilac

FIG. 8. The expectation from mating the double heterozygote to the double recessive; illustrating the 1:1:1:1 ratio.

cell, *bd*. Because of this, all of the genotypes can be written down in full, without the need for dashes to represent unknown genes.

It is possible to have another type of backcross when pairs of genes are involved. This is where one pair is heterozygous in both parents but the other pair is homozygous in one parent. Such a mating could be where one parent of genotype *BbDd* is mated to another of *Bbdd*. The expectations among the progeny are shown in Fig. 9. It will be seen that the same four phenotypes are produced, but in the ratios 3:3:1:1. The ratios are a combination of the 3:1 and 1:1 ratios discussed earlier, each gene pair assorting independently.

The advantage of making use of checkerboards is that the expectations are fully displayed in a systematic manner which can be checked for possible errors. The advantage, in fact, increases with the number of gene pairs. If the expectations were required for three pairs, their derivation would be difficult without the aid of a checkerboard. There would be eight different germ cells and a checkerboard of 64 small squares to portray all of the expected phenotypes. To illustrate such a cross, consider that a third gene has been added to the two previously described. Namely, short (*L*) versus long (*l*) hair.

Germ-cells from heterozygote *Bbdd*	Germ-cells from heterozygote *BbDd*: *BD*	*Bd*	*bD*	*bd*
Bd	*BBDd* Black	*BBdd* Blue	*BbDd* Black	*Bbdd* Blue
bd	*BbDd* Black	*Bbdd* Blue	*bbDd* Chocolate	*bbdd* Lilac

FIG. 9. The expectation for a mating in which one gene is assorting in a 3:1 and another in a 1:1 ratio; illustrating the 3:1:3:1 ratio.

As the above symbols imply, short is dominant to long hair. A mating between a blue short hair (*BBddLL*) and a chocolate long hair (*bbDDll*) will give a black short hair F_1 of genotypes (*BdDdLl*). Eight different germ cells will be produced by an F_1 of this genotype as follows: *BDL*, *BDl*, *BdL*, *Bdl*, *bDL*, *bDl*, *bdL* and *bdl*. The F_2 will be engendered from a random combination of these germ cells, yielding 64 individuals distributed in eight phenotypes:

No. of animals	Phenotype	Genotype
27	Black short hair	*B–D–L–*
9	Black long hair	*B–D–ll*
9	Blue short hair	*B–ddL–*
9	Chocolate short hair	*bbD–L–*
3	Blue long hair	*B–ddl–*
3	Chocolate long hair	*bbD–ll*
3	Lilac–grey short hair	*bbddL–*
1	Lilac–grey long hair	*bbddll*

The reader is urged to draw for himself or herself the checkerboard for the above mating and to confirm that the above expectations are correct. In doing so, he or she will probably agree that, while the completed checkerboard may look formidable, the arrangement of the entries are merely extensions of the steps indicated for the simpler checkerboards. The reader may like to try his or her hand with the checkerboard representing a mating of the F_1 *BdDdLl* with the triple recessive *bbddll*. This is a simpler checkerboard than the F_2, but should convince the reader that the genes do separate independently in the germ cells.

In fact, the whole of this section is an amplification of the second law of inheritance as stated in an earlier section of this chapter: namely, that for the independent assortment of characters in the individual.

Incomplete Dominance

Most pairs of genes are completely dominant or recessive to each other, so that the homozygous dominant or heterozygote are indistinguishable in appearance. In a number of instances, however, the heterozygote is phenotypically different from either homozygote. Genes of this type are said to display "incomplete dominance". An example is the merle gene *M*. The homozygote *MM* is nearly all white, has blue eyes and is usually deaf, while the heterozygote *Mm* has white markings on the head and shoulders and a dappled coat of normal and dilute (bluish, fawn) pigmentation. The normal gene *m*, from which *M* arose as a mutation, produces normal pigmentation.

The merle colour phase can produce a most attractive dog and the colour occurs in several breeds, in all probability the most well known being the Shetland Sheepdog. Merle mated to merle will produce white merle, merle and normal in a 1:2:1 ratio (normal meaning non-merle in this context, the actual colour depending upon other genes which may be present).

Germ-cells from merle *Mm*	*M*	*m*
M	*MM* White merle	*Mm* Merle
m	*Mm* Merle	*mm* Normal

(Row labels: Germ-cells from merle *Mm*)

FIG. 10. The expectation from mating two distinguishable heterozygotes; illustrating the 1:2:1 ratio.

Figure 10 shows the expectations diagrammatically. It is clear that the 1:2:1 ratio is a modified 3:1 ratio as a direct result of the three genotypes being distinguishable.

It is patent that the mating of merle with merle is not to be recommended if the breeding of the often blind and deaf white merle is to be avoided. If one wishes to breed merle dogs, the appropriate mating is to mate merle with normal, whence the expectation is half merle and half normal. This is in reality a backcross mating of the form *Mm* to *mm* to give *Mm* and *mm* pups in the ratio 1:1.

Multiple Alleles

If a normal gene can mutate once, it can do so again and again, even if the frequency of occurrence is very low. Each separate event could give rise to the same or to a different mutant. Over the years, a number of different mutants could be discovered, each one behaving as an alternative to the normal gene. The number of new alleles which could arise is large, but geneticists are wary of recognizing new alleles unless the phenotypic evidence is convincing. A sequence of mutants of the same normal gene is known as an "allelic series" or "multiple alleles".

As may be imagined, the number of known allelic series in the dog is quite small. However, one series is of considerable importance because it is responsible for several well-known colours. This is the agouti, so named because the different alleles have mutated from the agouti gene. Now, this is a basic gene in the genetics of mammalian coat colours. The designation of agouti is derived from a little rodent of this name with a nondescript grey-brown coat. The agouti gene is primarily responsible for this form of camouflaging coat colour. The corresponding colour of the canid family is the greyish-brown pattern of the wolf, or wolf-grey for short.

Allelic series are denoted by the distinctive and logical symbolism. Each gene and its alleles are represented by a different letter. For example, B–b for black versus chocolate mutant and D–d for dense versus dilute pigmentation. In the case of agouti, the appropriate symbols would be A–a, but with a modification to take account of more than one allele. These are given the same base letter, but with superscripts to denote the respective alleles. The main agouti alleles may be denoted as under:

Designation	Symbol	Example
Solid black	A^s	Black Labrador
Dominant yellow	A^y	Basenji
Agouti	A	Grey Elkhound
Saddle pattern	a^{sa}	Beagle
Tan pattern	a^t	Doberman

The normal gene is designated as agouti and is denoted simply as A. The solid black is due to an allele dominant to agouti, this aspect being indicated by a capital A but with the superscript s (from solid) to indicate the status of a mutant allele. For similar reasons, dominant yellow is denoted as A^y (y from yellow). On the other hand, the tan pattern allele is inherited as a recessive, hence this is indicated by a small letter a with the superscript t (from tan pattern). Another allele, saddle pattern (a^{sa}), is included in the above list, although there is ambiguity concerning its exact status. That a saddle allele exists is undoubted, but it could be merely a modified expression of agouti. This question will be dealt with fully in the subsequent chapter on the inheritance of colour.

The dominace relationships between the alleles is straightforward. Complete dominance seems to be the rule, in that each allele is dominant to all others below it in the series. That is, A^s to A^y to A to a^{sa} to a^t. A few people have opined that sable coloration, where the dominant yellow phenotypes have variable amount of black hairs on the head, shoulders and back, is due to heterozygosity for a^t, namely the genotype is $A^y a^t$ as opposed to $A^y A^y$. However, Little (1957) has stated that clear tan (yellow) Dachshunds have been found repeatedly to be of genotype $A^y a^t$. It is wise, therefore, to accept with caution the possibility of incomplete dominance of $A^y a^t$.

No matter how many alleles there may be in a series, only two will be present in an individual and only one transmitted in the germ cells. For example, a solid black heterozygous for dominant yellow or for tan pattern would be $A^s A^y$ or $A^s a^t$, respectively. It is not possible to have the genotype $A^s A^y a^t$ for instance. However, one particular mating could convey the impression that this is so, and the situation deserves to be explained. A black heterozygous for tan pattern ($A^s a^t$) mated to a yellow heterozygous for tan pattern ($A^y a^t$) would be expected to produce black, yellow and tan pattern in the ratio 2:1:1. If this seems curious, the details are shown by Fig. 11.

Germ-cells from black $A^s a^t$

Germ-cells from yellow $A^y a^t$

	A^s	a^t
A^y	$A^s A^y$ Black	$A^y a^t$ Yellow
a^t	$A^s a^t$ Black	$a^t a^t$ Tan pattern

FIG. 11. The expectation from a mating involving three alleles at the same locus; illustrating the 2:1:1 ration. Note that only two alleles can be present in the same individual.

The explanation resides in the random separation and recombination of the alleles, together with their dominance relationships. A^s is dominant to both A^y and a^t, producing two blacks, while A^y is dominant to a^t, producing one yellow.

Another feature of multiple alleles, which can mislead a novice, is that some results could imply that an individual is heterozygous for a certain allele when in actuality it is not. For example, a mating of solid black of genotype $a^s a^t$ to a yellow of genotype $A^y A^y$ will produce two sorts of puppies, namely black ($A^s A^y$) and yellow ($A^y a^t$). It could be imagined that the black parent is of genotype $A^s A^y$, but the assumption would be false. It is advisable to appreciate that an individual could be heterozygous for an allele lower in series than that appearing among the offspring. No ambiguity will arise, of course, if the heterozygote is mated to an individual homozygous for the lowest allele of the series.

Another important series of alleles is known as the extension. This has three known members as follows:

Designation	Symbol	Example
Brindle	E^{br}	Brindled Bulldog
Normal extension	E	Black Labrador
Non-extension (yellow)	e	Yellow Labrador

The series is called extension because the alleles are concerned with the extension of black pigment in the coat. The normal gene E produces the normal amount, as shown by the solid black or tan pattern. This gene has produced two known alleles: brindle (E^{br}), a dominant gene producing a brindled pattern of short vertical stripes of black pigment in the normally yellow areas of the coat, and non-extension (e), a yellow phenotype due to complete absence of black pigment. The e allele is often called "yellow" because of the phenotype. Actually, it may vary from yellow to light red, the variation being due to modifying genes (see later).

The importance of the agouti and extension series is that between them they control the development of black pigment in the coat. The respective alleles of the two series interact phenotypically with each other to produce most of the basic colours of dogs. These interactions will be taken up and discussed in a later chapter.

Mimic Genes

It is not uncommon for two (or more) independently inherited genes to display closely similar or identical phenotypes. This sort of effect is not due entirely to coincidence, but arises in the main from the fact that the expression of each character is controlled by many genes, any of which could mutate to produce an allele with similar end effects. Such genes are often termed "mimics", alluding to their similar phenotypes.

A most remarkable instance of mimicry occurs in the dog which has been the source of confusion in the past. The breeding results which usually leads to the discovery of mimics is when two apparently recessive individuals surprisingly produce offspring with the dominant phenotype. To anyone versed in elementary genetics, this result would seem to contradict one of the tenets of genetics, that animals showing recessive characters will breed true.

The situation of mimic genes in the dog is intriguing. It arises when a mating of two yellow dogs produces black pups. This result has been observed on several occasions, where mistakes in mating or errors of observation can be excluded. The most plausible explanation is that two yellow mutants have been unbeknowingly mated. The reason is the existence of mutant genes-producing yellow phenotypes. In symbols, this is what has happened: yellow or genotype $A^y A^y EE$ has been mated to a yellow of genotype $A^s A^s ee$, yielding black progeny $A^s A^y Ee$. The pups are black because A^s and E, introduced from different parents, are dominant to A^y and e, respectively.

It may be thought that such crosses will be unlikely; this is true, but the point is that they have occurred and will probably recur. Burns and Fraser (1966) describe and illustrate a mating of a sable Collie ($A^y a^t EE$) with a yellow labrador ($A^s A^s ee$) which produced a litter of five black ($A^s A^y Ee$ or

$A^s a^t Ee$) pups. Sable Collies have a suffusion of black pigment on the head and back, but genetically they are A^y. The suffusion of black pigment is usually a means of distinguishing A^y from *ee*, since the latter rarely (never?) display black pigment. On the other hand, many A^y dogs do not show black suffusion; these have been called "clear sables". These are the A^y yellows which can be confused with yellows of *ee* genotype. The depth of pigmentation of yellow is no guide to the separation of A^y and *e* animals, as this is due to independent genes.

Although a cross of the above nature is not likely to be carried forward into the F_2 generation, it will be informative to examine the expectation.

Germ-cells from black A^sA^yEe (columns); Germ-cells from black A^sA^yEe (rows)

	A^sE	A^yE	A^se	A^ye
A^sE	A^sA^sEE Black	A^sA^yEE Black	A^sA^sEe Black	A^sA^yEe Black
A^yE	A^sA^yEE Black	A^yA^yEE Yellow	A^sA^yEe Black	A^yA^yEe Yellow
A^se	A^sA^sEe Black	A^sA^yEe Black	A^sA^see Yellow	A^sA^yee Yellow
A^ye	A^sA^yEe Black	A^yA^yEe Yellow	A^sA^yee Yellow	A^yA^yee Yellow

FIG. 12. The expectation of mating black heterozygous for the two yellow genes: illustrating the 9:7 ratio.

The expectations are set forth in the checkerboard of Fig. 12. The expected ratio of black (A^s–E–) to yellow (A^yA^yE–, A^s–ee and A^yA^yee) animals is 9:7, which is a 9:3:3:1 ratio with the last three phenotypes added together. The reason is that all three are yellow in colour and are indistinguishable. Hence, for practical purposes they have to be treated as a single group.

Masking of Characters

Dominance is the term employed to describe the over-riding effect of one member of a pair of genes over that of the other. However, the effect of some genes are so over-riding or far-reaching that they are capable of masking the effects of independently inherited genes. This masking is known as "epistasis" to distinguish it from dominance. Unfortunately, dominance and epistasis are so similar in action that the two are often incorrectly used by breeders. Usually the term dominance is employed where epistasis would be more appropriate.

Easily the most remarkable instance of the effect of one gene masking that of others not allelic to it is that of recessive *e* upon the agouti series. The effect of *e* is to remove all black pigment from the hair, so that the coat becomes yellow as described in a previous section. This means that a yellow phenotype is produced regardless which agouti alleles are present. That is, in symbols, A^s–*ee*, A^y–*ee*, *A*–*ee* and $a^t a^t ee$ are all yellow in appearance. The reason is that the agouti alleles are concerned with the pattern of black pigment; remove this pigment and the agouti alleles cannot find expression.

Unless one is aware of the existence of epistasis, the results of certain matings could seem puzzling. For instance, the mating of a tan pattern ($a^t a^t EE$) with a yellow ($A^s A^s ee$) will produce black ($A^s a^t Ee$) puppies. It may be wondered, how can this be? Both tan pattern and non-extension yellow are recessive characters, yet here is the production of solid black, a dominant! The apparent paradox is cleared up once it is realized that the *ee* yellow is carrying A^s.

Should the F_1 of genotype $A^s a^t Ee$ be mated together, both tan pattern and non-extension yellow will reappear in the F_2. Furthermore, these phenotypes occur in a peculiar ratio which is characteristic of epistasis. The checkerboard of Fig. 13 provides details of the expectations for the F_2 generation. Bearing in mind the dominance and epistatic relationships between the alleles, the ratios of solid black, tan pattern and yellow occur as 9:3:4. These ratios are expected, because the three yellow genotypes $A^s A^s ee$, $A^s a^t ee$ and $a^t a^t ee$ are phenotypically indistinguishable. A little thought will reveal that the above ratio is but a modification of the ordinary 9:3:3:1.

The masking need not be complete: it could be incomplete, in which case it is usually possible to make an educated guess at the genotype. For instance, both A^y and *e* alleles are epistatic to *B* and *b* as regards coat colour. However, both A^y and *e* yellow animals possess pigmented nose leather and lips. This is black in *B*– individuals but liver in *bb*. The same distinction can be made for "black" yellows of genotypes A^y–*B*– or *B*–*ee* versus "liver" yellows of genotype A^y–*bb* or *bbee*, even if the coat is identically yellow. The iris of the eye is often a lighter shade of brown in the latter individuals. From a distance, too, they often appear "cleaner" or "brighter" in the quality or tone of the yellow because the residual black pigmented hairs of the coat (barely visible to the eye individually) are changed to chocolate, which blends in better with the yellow background. This effect is particularly noticeable for A^y yellow dogs.

Germ-cells from black A^sa^tEe

Germ-cells from black A^sa^tEe	A^sE	a^tE	A^se	a^te
A^sE	A^sA^sEE Black	A^sa^tEE Black	A^sA^sEe Black	A^sa^tEe Black
a^tE	A^sa^tEE Black	$^ta^tEE$ Tan pattern	A^sa^tEe Black	a^ta^tEe Tan pattern
A^se	A^sA^sEe Black	A^sa^tEe Black	A^sA^see Yellow	A^sa^tee Yellow
a^te	A^sa^tEe Black	a^ta^tEe Tan pattern	A^sa^tee Yellow	a^ta^tee Yellow

FIG. 13. The expectation of mating double heterozygous blacks; illustrating the 9:3:4 ratio.

Another interesting epistatic relationship occurs between A^s and E^{br} genes. If a mating is made between a black ($A^s A^s EE$) and a brindle $A^y A^y E^{br} E^{br}$, black offspring of genotype $A^s A^y E^{br} E$ will be produced. The F_2 generation will assort into three phenotypes, black, brindle and yellow in the ratios 12:3:1. Full details are shown in Fig. 14. The ratio may seem to be curious, but is due to the epistatic nature of A^s to both E^{br} and E. Both A^s–EE and A^s–E^{br}– individuals are phenotypically black.

In practical breeding, neither of the ratios 9:3:4 or 12:3:1 are likely to be encountered very often, but the ratios indicate that while the genes themselves may be inherited independently, it is possible for interactions to occur at the phenotypic level. The ratios provide an insight into the sort of interplay which can be expected to arise between genes from time to time.

A fourth example of masking is that of the extremely or all-white dogs. It may be diffiicult or impossible to be certain of the genes which may lie concealed under the white coat. If the individual has small coloured spots or patches, it may be possible to obtain some idea, depending upon their position. Thus, a black spot on the back would exclude the presence of dilute (d), but would not distinguish between solid black (A^s) or tan pattern a^t). However, a patch on the head might include an area which would be expected to be black if the dog carries A^s, but tan patterned if it carries a^t. Most white dogs have pigmented noses and lips, hence it is usually possible to decide if the animal is basically black (B) or chocolate (b) pigmented.

Germ-cells from black $A^sA^yE^{br}E$

Germ-cells from black $A^sA^yE^{br}E$	A^sE^{br}	A^sE	A^yE^{br}	A^yE
A^sE^{br}	$A^sA^sE^{br}E^{br}$ Black	$A^sA^sE^{br}E$ Black	$A^sA^yE^{br}E^{br}$ Black	$A^sA^yE^{br}E$ Black
A^sE	$A^sA^sE^{br}E$ Black	A^sA^sEE Black	$A^sA^yE^{br}E$ Black	A^sA^yEE Black
A^yE^{br}	$A^sA^yE^{br}E^{br}$ Black	$A^sA^yE^{br}E$ Black	$A^yA^yE^{br}E^{br}$ Brindle	$A^yA^yE^{br}E$ Brindle
A^yE	$A^sA^yE^{br}E$ Black	$A\,A^yE^{br}E$ Black	$A^yA^yE^{br}E$ Brindle	A^yA^yEE Yellow

FIG. 14. The expectation from mating double heterozygous blacks; illustrating the 12:3:1 ratio.

Expression and Penetrance of Genes

A comment that no two dogs are exactly alike is so commonplace as to be meaningless. Yet, has one ever stopped to consider the implications? The variation affects the phenotype of both the normal and mutant gene, but seemingly more particularly the latter. Some mutant phenotypes are relatively constant while others vary considerably. An example of the former is the solid black; nevertheless, if these are examined critically, variation is apparent in the intensity of black and in the number and/or distribution of yellow-grey hairs in the coat, lying under the all-black guard hairs. These latter contribute towards the expression between a coal-black and a brownish-black.

Another variable character is tan pattern, especially if one considers the tan areas of the pattern. These can vary in two features. The typical expression is that of tan areas on the stomach, inside of the front and back legs, across the chest and on the sides of the muzzle and face, with two spots above the eyes. These markings may be very restricted, occurring only on the stomach and legs, or extensive, so that the pattern approaches that of a dark saddle. In addition, and independently, the intensity of the yellow areas may vary from pale yellow to rich reddish tan. The variation shown by tan pattern is in fact an excellent example of variation of expression of a mutant gene.

A more complicated case is that of the expression of white spotting. Different breeds exhibit characteristic amounts of white, indicating that the average expression is controlled genetically, Examination of the distribution of white areas reveals a fairly orderly progression from patches on the chest and stomach, a collar around the neck and shoulders, to very extensive white on the body, all with much individual variation. Extreme-white animals are mostly white with smaller and smaller patches of coloured fur. Several allelic genes are probably responsible for the variation, but one in particular is worth mentioning. This is piebald (s^p), a mutant gene of non-spotted (S). The expression of this gene is so variable that $s^p s^p$ animals can have white markings ranging over the whole pied spectrum, except for extreme-white (these do occur, but are very exceptional).

The variation of white spotting is such that it can be portrayed as a curve on a graph by plotting the percentage of animals exhibiting successive degrees of white, as Little (1957) has shown for the white marking of Beagles (Fig. 15). This could be termed the "expressivity profile". The curve may be expected to differ between breeds and even between strains within a breed, indicating that the expressivity is under genetic control. This is a refined method of analysing the expression of a gene and could be used to show the variation of tan pattern if one so wished.

The point to emphasize is that the expression of a gene can vary; in some cases very little, in others very considerably. If a character is very variable, it is conceivable that some individuals could fail to express the character at all. This means that the individual would appear normal although the genotype is that of a mutant. This curious phenomenon is known as "impenetrance"; that is, the expression of the gene has failed to "penetrate". Fortunately, impenetrance is not an important factor in coat colour inheritance. The expression of a gene may vary, but its penetrance is usually 100 per cent.

As an example of impenetrances of a coat colour character, mention may be made of individuals with minor white spotting (toes, forelegs, breast or stomach spots, etc.) which sometimes crops up in breeds which should be entirely non-spotted. Their occurrence can be a nuisance. The positions and sizes of the white spots is variable, and it is probable that some individuals which ought to exhibit spots do not. Such individuals are called "normal overlaps" because of their apparent normal appearance. These will, however, breed according to their genotype and thus perpetuate the spotting. This is unfortunate for those breeders who are endeavouring to eliminate the blemish. The term normal overlap is not confined to minor white spotting, but is used for all individuals which should display a mutant phenotype but in fact are of normal appearance.

It is not uncommon for morphological anomalies to be variable in expression. That is, the severity of the anomaly will vary from dog to dog, even to the extent that individuals may fail to express the anomaly despite their mutant genotype. These individuals are normal overlaps in the full

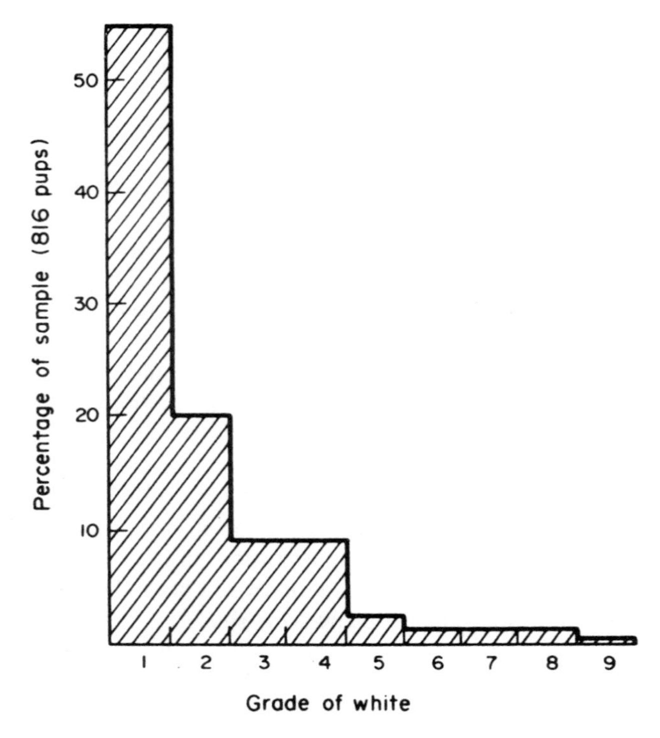

FIG. 15. The percentages of increasing amount of white for a large sample of Beagle pups; graded for amount of white as shown by Fig. 29.

meaning of the term. The reason why many genetic anomalies show impenetrance is that the development of the foetal dog is remarkably resilient. Whenever an organ is not developing properly, the growth processes act to rectify the defect. This is why so many anomalies are viable instead of dying at an early embryonic age. Unfortunately, the self-rectification may be incomplete and a defective pup is born. In other cases, when the self-rectification has been successful, an apparently normal pup is born. This may seem fine, but in those situations where attempts are in hand to eliminate the defect, these normal overlaps are a real nuisance, for the dog will pass on the genes for the defect to future generations. Unless the breeder is aware of the possibility of normal overlaps, he or she will be unware of what is happening.

Sex-linked Inheritance

The cat has the classical case of sex-linked inheritance in the form of the fascinating parti-coloured tortoiseshell breed. The colour only occurs in the female and is determined by a mutant gene on the *X* chromosome. The dog has no comparable gene. The brindle of the dog has merely a passing resemblance to tortoiseshell of the cat and occurs in both sexes, a fact which automatically rules out sex linkage.

Sex linkage occurs whenever a mutation occurs in the *X* chromosome. The most practical outcome of this event is that different results are to be expected according to whichever sex contributes the mutant in a mating. For example, Table 5 shows the expectations for the six possible matings involving the sex-linked orange or tortoiseshell colours of the cat. Should a mutant for sex-linked orange occur in the dog at some future date (as it may easily do), it would be inherited in an identical fashion. The tortoiseshell pattern occurs only in the female, because it is produced by the heterozygosity of the orange gene *O*. Only the female has two *X* chromosomes and can be heterozygous for *O*, i.e. possess the genotype *Oo*. The male with only one *X* is either orange (*OY*) or black (*oY*). The homozygous *OO* female is orange and phenotypically identical to the orange male. In deriving the results of Table 5, it is necessary to remember that the male has only one *X* chromosome and the *Y* chromosome takes no part in the inheritance of the *O* gene. The reader is urged to work out some of the expectations, using checkerboard diagrams to ensure accuracy.

TABLE 5

Expectations for Matings involving a Sex-Linked Colour Gene; using the Orange Gene of the Cat as an Example

Mating		Offspring	
Dam	Sire	Males	Females
Orange	Orange	Orange	Orange
Black	Orange	Black	Tortoiseshell
Tortoiseshell	Orange	Orange	Orange
		Black	Tortoiseshell
Tortoiseshell	Black	Orange	Tortoiseshell
		Black	Black
Orange	Black	Orange	Tortoiseshell
Black	Black	Black	Black

If the dog does not possess any coat colour sex-linked genes, the animal is infamous for having haemophilia, a similar, if not identical, blood disease as occurs in man. In dogs, as in man, the disease is rare but recurrent, for it has been reported in many breeds over the years. The interesting aspect is that it is due to a mutant gene on the *X* chromosome. The hallmark of the disease is that haemophiliac males are born to normal parents.

The genetic situation is shown by Fig. 16. The mating is equivalent to that of the fourth row of Table 5.

Note that the male has no need to be related to the female. None of the female but half of the male pups are to be expected to be haemophiliac. Half of the females, however, will be heterozygous for the haemophiliac gene, and this is the usual means by which the mutant is handed on from generation to generation. It might seem strange at first that the normal brothers of affected males are completely devoid of the gene. It is the

Germ-cells from female carrier $X^H X^h$ \ Germ-cells from dog $X^H Y$	X^H	Y
X^H	$X^H X^H$ Normal	$X^H Y$ Normal
X^h	$X^H X^h$ Normal	$X^h Y$ Haemo-philiac

FIG. 16. Inheritance of haemophilia: illustrating one form of sex-linked heredity.

apparently normal and harmless looking heterozygous sisters which will produce haemophiliac sons, even if the breeder takes care to mate with an unrelated male. This procedure would be correct for ordinary (non-sex-linked) recessive characters, but not for sex-linked.

Sex-limited Expression

There is another type of sex associated heredity which, unfortunately, is sometimes mistaken for sex linkage. This is where the expression of a gene is confined to one sex, although the gene itself is borne by one of the ordinary chromosomes. This is sex-limited inheritance. It is possible to have partial sex-limited inheritance where the expression of a gene is more obvious or comes through more strongly in one sex than the other. No clear-cut case of sex-limited expression affecting coat colour is known for the dog, but other important features are affected in this manner.

The various sex characteristics are determined in the first instance by the sex chromosomes, but this does not indicate that these are sex-linked in heredity. The ability of a bitch to produce large litters and to rear these in a satisfactory manner is determined by genes promoting high fecundity, abundant and nutritious milk flow and strong maternal drive. These are inherited by genes common to both sexes, although they can only find expression in the female. Thus, a bitch deficient in one of the above features, due to genetic causes, will be passing the deficiency on to both her sons and daughters, not the daughters alone, as might be imagined. All of the offspring of such a bitch must be regarded with equal suspicion.

Similar reasoning also applies to the male in respect to sexual virility which all stud dogs should possess to a high degree. This feature cannot be shown by bitches, but they can transmit the ability or inability to their sons. Because of the importance of the stud dog in breeding, all males destined to become stud animals should be chosen from large litters produced by proven mothers sired by fully competent males. This recommendation may seem idealistic (if commonplace), but it must be remembered that most strains of dogs are subject to some degree of inbreeding. The two features most affected by breeding are body size and fecundity, and it is wise to select for high fecundity at all times to counteract the possible insidious effects of inbreeding.

An anomaly which obviously has sex-limited expression is cryptorchidism. Undescended testes is a defect which can only afflict males. It is probable that cryptorchidism (whether unilateral or bilateral) is determined in part by heredity. Females may not be able to exhibit the defect, but they could carry liability for it and pass the liability to their sons. Animals of both sexes of any strain with a history of cryptorchidism should be regarded as equally liable; not merely the males, as may be assumed upon a simple view of the topic. Liability in this context may be defined as the ability to pass on the defect to future generations.

The sex difference in size and weight is an example of partial sex-limited expression of polygenes. There is no question that the sex difference of either feature is due to sex-limited genes. Both sexes receive the same genes, but body growth differs between the sexes. This is mediated by the sex hormones and is a secondary sex characteristic. It makes little difference whether one considers that the growth potential is slightly inhibited in the female or that growth is accelerated slightly in the male; in the event, males surpass the females. The difference is clearly on a percentage basis, so that the size differential is greater for large breeds than for small. There are, however, subtle differences of relative size between breeds, affecting some features more than others, nevertheless, the influence of sex is still evident.

Genetic Linkage

Each chromosome is composed of thousands of genes, each one capable of mutation. By chance, mutant genes could arise from different normal genes on the same chromosome. Should this happen, the implication of independent separation in the germ cells will no longer be true, because it is independent assortment of the chromosomes at meiosis which guarantees this. In practice, two genes entering the mating from different parents will fail to recombine as freely as expected; or, if entering the mating from the same parent, will fail to separate as often as expected. This effect of staying together is known as "linkage" and the two genes are said to be "linked". It implies, in effect, that the two genes are on the same chromosome.

None of the mutants described in this book are known to be linked (with the exception of sex-linked genes, a special case). This is not due to tests for linkage which have proved to be negative, but from a complete absence of such tests! It is traditional to assume that all mutants are inherited independently until the evidence for linkage is incontestable. There is little prospect of this occurring in the dog, since the animal is so rarely employed in genetic studies likely to lead to the discovery of linkage.

There is reason to believe that linkage will be uncommon in the dog. In part, this is due to the small number of mutant alleles with which breeders normally have to deal, but also to the large number of chromosome pairs (39) for the species. The larger the number of pairs, the less likely that two mutants will reside in the same chromosome. This should not mean that breeders should ignore the phenomenon completely. They should be aware of it, but treat the possibility as a rather remote one until proven otherwise.

It may be wondered: if two alleles occur on the same chromosome, would these not stay together permanently? How is it that some recombination occurs, even if this is less than that free recombination? The answer lies in one of the subtleties of meiosis. When the chromosome pairs come together in the middle of the cell, the pairing is so intimate that lengths of chromosome are in close contact. During the process of pairing, internal stresses cause the chromosome to break. In a proportion of breaks, the ends of the same chromosome do not rejoin, but join up with the partner chromosome. In this manner, the chromosome pairs have exchanged segments with each other. The breaks occur at random along the chromosome; implying that unless two genes are sited very close together these will be separated eventually. In fact, the greater the distances apart are two genes on the chromosome the greater the chances that they will be separated. The degree of separation can be demonstrated in precise experiments. The point is that unless two genes are extremely closely sited on the chromosome (close linkage is the term used), they will be recombined eventually. The position a gene occupies upon a chromosome is known as the "locus" of that gene. The symbol given to a locus is the same as that for the basic gene. When more than one locus is discussed, they are referred to as loci. Therefore, the loci of the black and dilution genes are *B* and *D*, respectively.

Linkage is an important aspect of genetics and most breeders, if they read widely, will doubtless encounter the term and the possible effects it may produce. For this reason it is desirable to consider the topic, if briefly.

It is not an aspect which is likely to trouble the thoughts of many breeders, but those who wish to learn more are advised to consult a general gentic textbook. The implication of linkage are identical for all mammals and, even if the textbooks do not mention the dog specifically, no misleading ideas should be picked up. The most obvious effect of linkage in practical breeding is that the genes do not recombine as freely as one would expect.

Continuous Variations

Broadly speaking, there are two sorts of heredity. That which deals with sharply defined alternatives, such as black versus chocolate pigment or dense versus dilute pigmentation as discussed earlier, and that which deals with characters varying smoothly from one extreme to the other, such as the intensity of yellow pigmentation or body size. The first sort is spoken of as "discontinuous" and the second as "continuous".

Could this continuous variation be inherited and, if so, what is the nature of the genes involved? The answer to the first part of the question is yes, a sizeable part of the variation is inherited, although the proportion may vary from character to character. In this respect, continuous variation differs from the discontinuous. In the latter, the size of the step between alternatives tends to swamp the incidental background variation. For instance, for the black-chocolate pair of characters, the black may vary in colour, as may the chocolate, yet it is not possible to confuse black with chocolate. Both retain their separate identities and are inherited independently. True, some of the blacks tend to brownish and some chocolate can be exceptionally dark, yet there is a distinct discontinuity from one colour to the other.

The situation is different for a continuous character. There are no abrupt changes in the transition from one extreme to the other. If this is so, how can one be sure that part of the variation is inherited? One method is by observing the results of selective breeding for extremes of expression. If the differences in body size, say, are not inherited, it would be impossible to produce small or large animals by selective breeding. This has been undertaken on several occasions with quick-breeding mice, and both small and large strains have been successfully established. This is proof that at least part, probably the larger part, of the variation of body size is inherited.

The dog, however, offers sound evidence of a different sort that body size is inherited. This is the fact that many differences of size exists between breeds. Size and conformation are major constituents of breed differences. What is more, the breeds are expected to breed true for their respective sizes. None of this would be possible if body size is not largely determined by heredity. Dog breeders have in effect duplicated over many decades of selective breeding the same results accomplished in a year or two with mice. They have, in fact, achieved something more, for dogs are, relatively, very variable for body size, and breeders have shown that it is possible to stabilize intermediate levels of body size. This is a more difficult task than selection for extremes, but one which can be achieved by persistent selection and inbreeding.

Body size is a typical continuous character and it is determined by many genes, each one having a small effect individually but acting co-operatively are capable of producing remarkable variation. It is possible, therefore, to classify character upon the basis of the number of genes which determine them. Thus, the chocolate or dilute colours are produced by a single mutant gene. They are termed "monogenic" characters, "mono" meaning one. Body size, or variation in intensity of yellow pigment, on the other hand, is produced by many genes. These characters are termed "polygenic", "poly" meaning many.

The concept of character being determined by many genes is important in genetics, so important, in fact, that the term "polygenes" has come to mean the actual genes involved in the production of a continuous character. Thus, one speaks of polygenes as a group, for body size or for intensity of yellow pigment. In other words, labelling different groups by the character they engender. More shortly, they can be referred to as size polygenes or yellow intensity polygenes. Another, less technical, term is to call those genes with large effects "major" genes and those with small effects "minor" genes.

A polygenic character, therefore, is one which varies continuously and is produced by many genes with similar effects, although the effect of each one is small relative to the total variation. Though the action of each polygene is small, the effects are cumulative, so that an individual which has a number is phenotypically different from one which has only a few. Because numerous polygenes are involved, the inheritance of continuous characters is seemingly blending. The polygenes themselves do not blend, but assort in the manner of genes with large effects. It is the phenotype which shows blending, in that the offspring are generally intermediate to that of the parents. The polygenes must assort independently because they are borne by the chromosomes, but because their individual effects are so small, they produce a blending phenotype.

It is conventional to regard polygenes as having plus or minus effects with respect to a character. Those which increase the expression of a character are the plus polygenes, while those which decrease are minus. All polygenes are inherited in pairs, so the plus and minus designation apply to one of the pair and, collectively, to groups acting in the same direction. In the case of body size, the plus polygenes are clearly those producing increments of size, while the minus polygenes are those having the opposite effect. That both types of polygenes exist is shown by the very large and exceedingly small breeds of dog. The extremely large breeds have a genotype of plus polygenes, while the small breeds have a genotype of minus polygenes. It is impossible to say how many plus or minus polygenes are

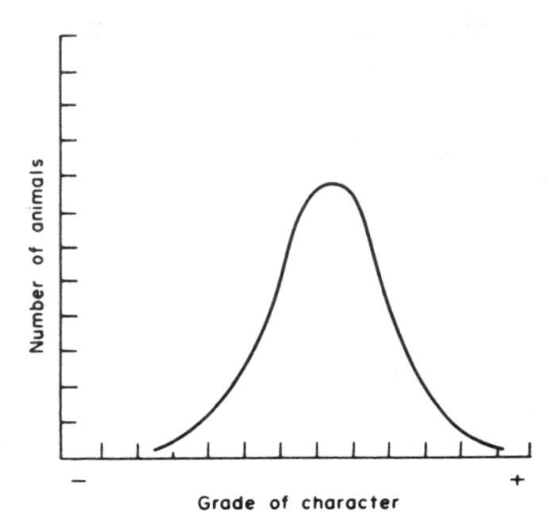

FIG. 17. The sort of idealized curve to illustrate the number of animals of each grade of a character determined by polygenes. Some characters, such as weight or size, will show a lopsided curve, but the principle is the same.

concerned in the heredity of body size beyond that of saying that there are many rather than few.

In addition to general size polygenes there are several groups concerned with conformation. These are distinct from, but interact with, those responsible for sheer size. The length of the long leg bones is one example, since it is possible to have dogs with varying limb lengths for similar body sizes. The two groups interact in the sense that the overall patterns of growth will not permit absurd combinations, although it can produce the odd strange-looking individual. Skull shape is governed by another group of polygenes in that one can have broad as well as slim skulls, but again there are constraints. Excessively broad or slim skulls are only found with stocky or slim body conformation, respectively. This is to imply that there are yet other polygenes controlling body conformation as distinct from sheer size. These considerations indicate two items; one, that groups of polygenes can often be subdivided, as different components of a polygenic character are analysed, and two, that size and conformation is a complex character, capable of subdivision into simpler components. These components are polygenic determined and can be manipulated by selective breeding. It is possible to have breeds of similar size or weight but of different conformation or, for that matter, the reverse.

Three groups of polygenes affecting coat colour merit a few paragraphs at this time, if only for purposes of definition. The first are the rufus groups which are responsible for the intensity of yellow pigmentation. These can range from lemon-yellow to rich red, with many intermediate grades, the differences being due to polygenes. The variation of intensity of the yellow in saddle and tan pattern animals is almost certainly due to the same group. This, of course, implies that the group is widespread and contributes substantially to the ideal colour of many breeds.

Another group is responsible for the amount of black hairs which change the A^y yellows into a sable. A single gene could be responsible for sable, but this is doubtful because of the variation. Even if a single gene is involved, an explanation is still required for the variation of expression. The relevant polygenes are known as umbrous. Most A^y yellows have dark hairs, usually on the head and a faint spinal strip. As the number of dark hairs increases, the sable is produced, with various degrees of suffusion, especially on the head, shoulders and back.

The saddle pattern is subject to considerable variation. The extremes are a lightly pigmented saddle, restricted in area, to a dark saddle which has spread over most of the sides of the body. The saddle pattern is determined by a major gene, but the expression is strongly influenced by polygenes. It is possible that the same polygenes (umbrous group) may be involved in darkening both the sable and the dark grades of saddle, but this is pure conjecture at present. This aspect is mentioned because it is unwise to postulate more groups of polygenes than is absolutely necessary.

Manipulation of polygenes cannot be accomplished in the same easy manner as major genes. It is easy to follow the transmission of the latter by inspection of the phenotype and knowledge of the laws of inheritance. It is impossible to do the same with individual polygenes, but the position is not hopeless. Groups of polygenes can be manipulated, although different techniques have to be employed. These can be summed up under two headings, selection and systems of mating. Both of these topics are discussed in the next chapter.

Modifying Genes or Modifiers

Polygenic variation affects all characters to a greater or lesser degree and is present in all genetic situations. It has also been written about from different points of view. This is particularly true for those polygenes known as modifying genes or "modifiers". This is a term to describe those groups which appear to modify the expression of one of a pair of major genes but not (apparently) the expression of the other.

The variation of blue colour should illustrate the point. The intensity of blue varies over an appreciable range to experienced eyes, for it is possible to have dark, medium or light shades, all of which may be bred from black parents (one or both) as ordinary assortment of the *D* and *d* alleles. However, although the black pups may appear much the same, there may be distinct differences of shade between the blues. The variation of blue is polygenic in inheritance, and the group of polygenes have been termed dilution modifiers (but see later).

A similar picture emerges for chocolate; here, too, variation exists for intensity of the colour and is polygenically controlled. Black and chocolate pups may occur in the same litter (or to similar parents), but while the chocolates may vary between themselves, the blacks do not. One could speak of chocolate modifiers in exactly the same manner as dilution modifiers. It is probable that the two groups of modifiers are identical (at least for practical breeding purposes), the action of the polygenes being to darken (plus) or lighten (minus) the colour. They may even be part of the rufus group, which is known to be highly influential in determining the intensity of yellow pigmentation.

It is clear that the polygenes involved in the variation of blue or chocolate are independent of the *d* and *b* alleles. These two alleles simply provide the background for the variation to become apparent. It is very likely that the same polygenes modify the intensity of black colour (since the colour does vary) but to a much lesser extent, as observed by the human eye. It is easy to judge if a blue or chocolate dog possesses plus or minus polygenes by inspection of the intensity of the colour, but it is more difficult to do the same for a black individual. Black dogs undoubtedly carry the plus or minus polygenes, however, otherwise they would not produce dark or light blue pups on occasion.

The length of hair in dogs with long hair varies considerably, and is due to modifying polygenes in conjunction with the recessive long-haired gene *l*. The length of hairs in the normal short coat is also variable, but less so than in the long hair. It is easy to conject what is happening in this case. Suppose the long hair gene *l* increases hair length by a factor of 4. Any group of polygenes which increased normal hair length by x would be expected to increase the hair in long-haired individuals by $4x$. That is, the effects of the hair length polygenes are greatly magnified in long-haired animals. No wonder the variation is so apparent. No wonder, too, is the action of the polygenes so strikingly dependent upon the prescence of *l* and are referred to as modifiers rather than as polygenes for hair length in their own right.

Almost any character of the dog which can be seriously analysed will be found to vary, and at least part of the variation will be due to polygenes. It should be appreciated that not all of the variation will be genetic. The non-genetic part will arise from a multitude of causes, temperature, feeding, husbandry, uterine conditions, attention of the mother, almost any factor one may care to name, depending upon the character or characters under study. In the case of coat pigment or hair length, the role of non-genetic factors will be small.

An exception in this connection is white pattern, where the non-genetic component could be appreciable, as witness the pattern variation still shown by many breeds with white spotting. It is particularly interesting in that most of the variation cannot be attributed to tangible environmental factors. Much of the variation arises from chance development of the pigment cells (melanoblasts) during the development of the foetal pup in the uterus. Even so, little of it can be pin-pointed to particular aspects of the uterine environment. The non-genetic part of the variation appears to stem from the chance developmental processes within the foetus.

A character, or rather a collection of characters, which can be easily influenced by the environment are size, weight or conformation. The implicated factors are the feeding regime, particularly the diet, living conditions, particularly warmth and shelter, and exercise. The effects of these factors are largely negative. They can inhibit the full potential of an excellent dog, but have little effect on increasing the potential beyond that which the animal would reach naturally. All aspiring breeders should ensure that his or her dogs are well-cared-for and the kennels are properly maintained. Disease, too, can affect characters of this nature. Most infectious diseases can be checked nowadays, but may have an untoward effect upon the development of a promising young puppy.

Therefore, to summarize, continuous variation is the sum of three factors: polygenic heredity, environmental influences and chance development. The relative contribution of any one of these to the total variation will vary according to the character. Even the breed or strain is important, for a characteristic in one strain of dogs may be more strongly affected by one or another of the factors than in another strain. This aspect explains in part how successful results can be obtained with some strains but fail with others.

Threshold Characters

Polygenic inheritance is usually associated with characters which vary smoothly from one extreme of expression to the other. Any discontinuity which may be present can usually be ascribed to the assortment of a major gene. Such a character would be hair length, as discussed in a previous section. Hair length varies in short and long-coated dogs, but the two ranges of variation do not overlap because the step from one to the other is mediated by the *L/l* gene pair. The genetic situation can be summarized by saying that hair length is determined by a gene *l*, with large effects on hair length, and a group of hair length polygenes which interact with *l*.

However, it is possible for a group of polygenes to be associated with a discontinuity or jump in expression of a character. At first blush, it may be thought that a major gene is involved, but this possibility is not supported by a clear-cut 1:1 or 3:1 ratio of assorting alleles. The simplest explanation is that of a build-up of plus polygenes which enables a physiological process to be carried past a "developmental threshold", so that a new mode of expression can appear. The transition is abrupt and

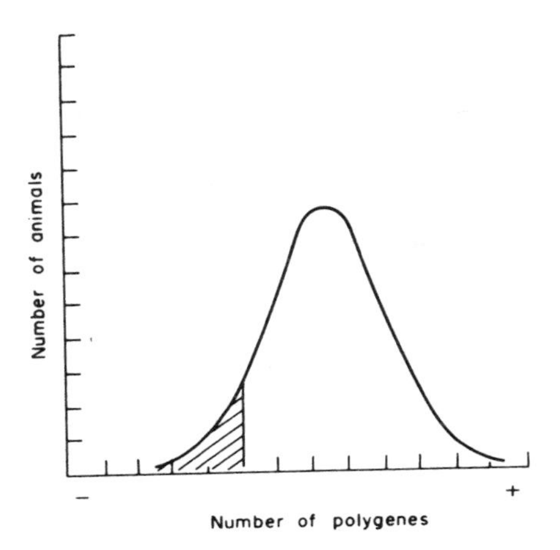

FIG. 18. The concept of the threshold character. Those animals which do not possess a minimum number of +polygenes fail to complete normal development and will develop into an anomaly (shaded area of the idealized curve shows the percentage in the total population).

characters determined in this manner are known as "threshold characters".

Another, and probably better, concept is to imagine that a development process requires a certain number of plus polygenes for successful completion; any number less than the optimum results not merely in decreased expression of the character, but one which is quite different. This suggests that most cases of threshold heredity will be associated with anomalies, and this is the situation. There are grounds for concluding that several recurring anomalies are due to polygenes interacting with a physiological threshold. This is the reason why this mode of inheritance must be considered, if only briefly. Breeders should be aware of the possibility and how to deal with it.

Any anomaly which occurs at a low frequency could be due to an uncommon recessive gene or by a threshold character. If the anomaly is due to a recessive gene, careful analysis of its occurrence usually reveals regular and predictable heredity. On the other hand, if the anomaly is a threshold character, the analysis usually fails to establish the ratios expected of a recessive gene. The observed ratio may be higher or lower than expected, usually lower, sometimes appreciably lower. Also, there is irregular inheritance, anomalous pups are produced by normal parents and normal pups from anomalous parents (if the anomaly is not too severe, so that the afflicted dog is viable and acceptable). Typically, it is the low incidence of occurrence of the anomaly in one breed or strain but higher incidences in others which is the first indication of threshold heredity.

In principle, an anomaly due to a threshold situation should be treated in exactly the same manner as those due to a recessive gene. That is, by refraining from breeding from any individual which has it (if the affliction is mild) or from known heterozygotes (if the affliction is serious). However, more drastic measures may be required. The testing of suspected heterozygotes (often feasible for anomalies due to recessive genes) may now be impracticable. If the anomaly is regarded as serious, selection has to be stringent, with all individuals which have produced anomalous young being prevented from further breeding, if one is determined to eliminate the anomaly as quickly as possible. If the anomaly is only regarded as a nuisance, it is possible to consider each case more lightly and balance the worth of the individual in other respects against the perpetuation of the anomaly.

Expectations and Chance

The expectation for various matings are expressed in terms of ratios, 3:1, 1:1 and others. This is purely conventional, for the same expectations could be expressed as proportions or as percentages. One could say, for example, that in the F_2, 75 per cent of the progeny will display the dominant character and 25 per cent the recessive, and in the backcross to the recessive 50 per cent will be dominants and 50 per cent will be recessives. This is merely a matter of convenience how the expectations are expressed.

However, there is another aspect that should be appreciated, and this is the chance factor. Consider the tossing of a coin. No one would doubt that the expected ratio of head:tails is 1:1. A few people might wonder why an exact 1:1 ratio is not realized after a series of tosses, while the majority would probably accept the outcome, provided the observed ratio is close to that expected. The reason why exact ratios are not obtained is due to chance, a factor which cannot be controlled. However, there is a rule of numbers which states that the larger the number of tosses, the closer will the observed ratio become to that expected. With small numbers of tosses, departures from the expected ratio must be accepted as inevitable.

Similar reasoning applies to the genetic ratios, whether these be 3:1, 1:1 or any other. In an F_2 litter of 8 pups, the theoretical expectation would be 6 dominants and 2 recessives, but these numbers will almost certainly not be obtained. In fact, a litter of 8 could easily be composed entirely of dominant animals, simply as a matter of chance. Over a series of litters, however, it will be found that other litters will contain more recessives than expected; so that, over the series, the ratio for the total number of animals will not differ too much from expectation. The fact that exact ratios cannot always be obtained with small numbers does not invalidate the usefulness of the expectations. These are invaluable in the planning of matings and in the development of breeding programmes. Animal breeding is very much a question of relative probabilities, an aspect which some people find exciting but others probably find frustating.

The question of probabilities will crop up again in another connection. A series of litters may be bred from a mixed group of dogs, some of which are homozygotes while others are heterozygotes. Those litters which contain pups of the recessive type are clearly bred from heterozygotes. However, if the number of recessives are totalled, the percentage may seem to be in excess of the expected 50 or 25 per cent. The reason is that some litters from heterozygotes will consist entirely of dominant pups, and these will not be included among the offspring from known heterozygotes. The chances of this happening will vary with the size of litter, being large for small litters but negligible for large litters. It is easy to correct for this element of chance, and the expectations for successive litter sizes are shown in Table 13, p. 192.

The two columns of the table represent the expected number of pups, with the recessive phenotype for matings of dominants with recessives (back-cross, first column) or dominants with each other (F_2), second column), litters with only recessive offspring being counted. An example will illustrate how the figures of the table should be employed. Suppose that matings are made between black (dominant) dogs which are producing occasional blue (recessive) pups. Six litters gives a total expectation of 9.4 which is quite close to the 11 blue pups of the example.

The calculation of probabilities can be important for another situation. A problem which often arises is whether a reputedly homozygous animal may safely be regarded as such. In principle, the solution is obvious. Suppose a black dog is thought to be heterozygous for blue dilution and it is desirous to check on the possibility. The procedure would be to mate the animal to a blue. If a single blue pup is bred, regardless of the number of black siblings, the animal is a heterozygote. However, suppose only black pups are produced, how safe would be the conclusion that the dog is a homozygote?

The difficulty is that a heterozygous dog may produce a litter of black pups by chance in a similar manner as a succession of coin tosses may result in a run of heads (or tails). It is obvious for the latter that equal numbers of heads and tails are expected, but this does not prevent a series of either heads or tails to occur occasionally. In a similar manner, only black progeny may be bred when equal numbers of black and blue are to be expected. However, the chances of this event occurring decreases with the number of pups. The problem is overcome by stipulating that more than a certain number of black pups are born before it is accepted that the individual is homozygous. In this way, the risk of being wrong (the error, as it is termed) can be made as small as desired.

For instance, suppose 5 young are born, all black, from a mating of a suspected black dog to a blue; the risk of error is 3 per cent. This may not be regarded as sufficiently low if the establishment of homozygosity is held to be important. The breeding of 7 black offspring would reduce the error to about 1 per cent, a more satisfactory level. This is equivalent to mating 100 heterozygous black dogs to a blue, rearing 7 pups from each, and discovering that one animal has produced an all-black litter.

Column A of Table 15 (p. 198) gives the error percentages for various numbers of progeny for backcross matings, all of which have shown the dominant phenotype. Returning to the above example, suppose a second litter of 4 black pups is born under the same conditions. These may be added to the 5 of the first litter, giving a total of 9 pups. Referring to the table (column A) shows that the risk of error is 0.2 per cent. This corresponds to a 1 in 500 level of error, and many people would regard this as being sufficiently low. However, the usual and more stringent test is to insist on 10 dominant progeny, in which case the error is as low as 0.1 per cent or 1 in 1000.

The situation may arise where it is impossible to mate the dog to be tested with the recessive. However, it may still be tested if other heterozygotes are available. The drawback with these matings is the larger number of offspring which must be examined for the same risk of error. At least 12 and 16 offspring are required to reduce the error to 3 and 1 per cent, respectively; for the 0.1 per cent level, the number is as great as 24 offspring. Column B of Table 15 gives the error levels for various numbers of

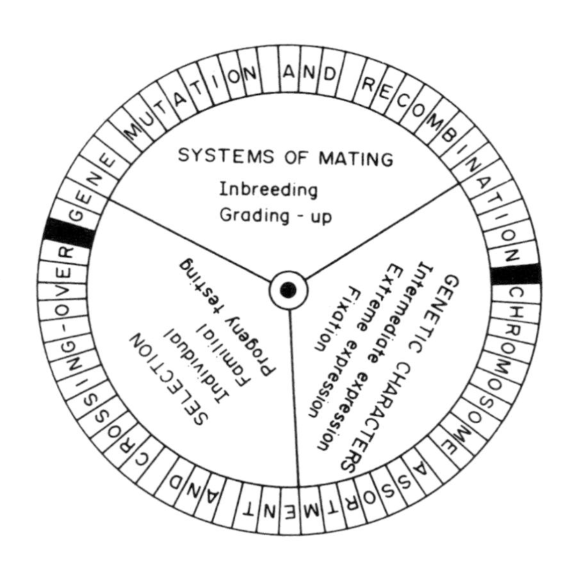

FIG. 19. The important factors of animal breeding shown as a wheel of chance. The breeder's art resides in the skilful manipulation of these to bring together genes which engender the ideal dog.

off-spring, all of which must be of the dominant phenotype. Remember that in these tests, if a single pup is bred of the recessive phenotype, regardless of the number of dominants, the individual under test emerges as a proven heterozygote. The use of the table may be illustrated by an example. Suppose 9 offspring are bred from a black suspected of being heterozygous for blue. Column B of the table for 9 offspring gives a percentage error of 7.5. This is much higher than the 0.2 per cent error for the same number of offspring produced by matings with a blue.

It may be thought that this is making an unnecessary fuss over the detection of heterozygotes (or "carriers" in breeders' terms) for a harmless coat colour gene. However, some breeders do, in fact, feel concerned if their charges are believed to carry certain recessive colour genes, but, nevertheless, the point is well taken. The problem discussed here has greater relevance for the detection of carriers of harmful recessive genes, an aspect which is discussed in a later chapter. This section is concerned with the principle underlying tests for unwanted genes, using an ordinary colour gene as the example.

5.

Selection and Breeding Policy

THE BASIC tenets of animal breeding are firmly rooted in genetic principles and the two topics cannot be sensibly separated. Indeed most books on genetics devote a chapter or two to animal breeding practice. However, the subject is worthy of serious study in its own right. This chapter presents a general outline, perhaps lightly in places but more deeply in others. Those who wish to extend their reading should consult Bogart (1959) and Hutt (1964). One of the finest books ever written on animal improvement is that of Lush (1945). Lerner's (1958) treatise is more difficult to comprehend but absorbing in revealing just how complex the subject can be if one endeavours to take account of almost every conceivable factor.

Some Preliminary Thoughts

The art and science of dog breeding rests upon two pillars of knowledge, namely selection procedures and systems of mating. In other words, the choice of certain individuals for the next generation and how these should be mated to each other. Elementary, one may say, but there are numerous ways of setting about these tasks, some of greater value than others and some of greater usefulness in solving certain problems. The two processes are complementary, but if their effects are to be appreciated, each must be considered alone in the first instance.

The phenotype of the individual is composed of a large number of characters of varying expression. The ideal dog is one in which the expression of each character is appropriately balanced in terms of the standard of excellence for the breed. This implies that intermediate grades of expression will be required for some characters but more extreme expression for others. The expression of genes is mainly determined by selective breeding, of which there are several kinds. The simplest is choice of the individual, but also selection can be between whole families (familial selection) or be based upon progeny testing. Reinforcing the selection are systems of mating, which embraces intense, moderate, weak or avoidance of inbreeding, the mating of like-to-like and the process of grading-up to improve inferior animals. These aspects are summarized by the diagram of Fig. 19. These are the tools which the breeder may use at his discretion.

Selective breeding is of considerable potency, whether this be straighforward selection or be one of the more sophisticated forms. One has only to survey the evolution and prolification of breeds of dogs over the last 100 years to appreciate this fact. Not only have many new breeds made an appearance, but the quality of almost all breeds has improved. Despite the efficacy of selection, this alone is insufficient. The different breeds must breed true to type to be worthy of the name. It might be imagined that the selective breeding is responsible, but it is doubtful if this is true in other than a very limited extent.

The degree of pure breeding is known as the "homozygosity". The homozygosity varies from zero to near zero (for example, in the mongrel pups from a cross between two breeds) to near 100 per cent for a highly inbred strain of pedigree dogs. It will be appreciated that it is virtually impossible to create 100 per cent homozygotic dogs, but any value over 80 or 90 per cent may be looked upon as being highly inbred. The converse to homozygosity is "heterozygosity", the proportion of impureness in an individual or breed. It is often more convenient to speak of the proportion of heterozygosity than the homozygosity.

Selection by itself is not very effective in bringing about a high degree of homozygosity. Inbreeding is much more consistent. Not necessarily close inbreeding, but the moderate form which results from mating of not too distantly related animals. Some inbreeding is certainly desirable, if not essential, during the formation of a breed in order that the breed characteristics should stabilize around a definite type. Those individual dogs which deviate too much from the ideal are either not bred from or are mated to a typical specimen of the breed. In general, a breed stabilizes itself and moves forward as an entity because the majority of animals tend to be descended from, or are bred to, a small group of élite individuals commonly regarded as the most outstanding. It is not uncommon for the élite group to be more interrelated than the ordinary members of the breed. In summary, breeders who have worked to produce the top-flight animals are exceedingly choosy in their mating, and hard-won experience has taught them the inadvisability of straying too far from the basic blood-lines of the breed.

Selection in Practice

Selection of animals for breeding can take a remarkable number of forms. Conceptionally, they all share the common principle of raising the general level of the breed. It is the method by which the selection is performed that differs, and this is where the skill of the breeder is taxed to the full in choosing the method to suit his particular problem. This can be a far from easy task, for any chosen method must fit in with the overall picture. Most breeders have a fair idea of the meaning of selection, but possibly only a vague notion of the kinds of selection which can be employed.

An important aspect of selection is its strength; the "selection intensity" as it is termed. It is not easy to gauge the intensity for all characters, but a most practical measure is that of calculating the proportion of pups retained for breeding. For example, a series of litters from three bitches could produce 21 pups, of which it is decided to keep 3 pups for future breeding. This represents 3/21 = 1/7th or about 14 per cent of the total number of pups. Now suppose a second round of litters were produced from the bitches, totallying 19 pups on this occasion. It is still decided to retain 3 pups, so the second batch of 19 is compared with the 3 from the first litter, and the best 3 chosen. Now, this represents 3/(21 + 19) = 3/40th or about 8 per cent.

It should be obvious from these calculations that, provided the better pups are always chosen, the smaller the percentage, the better the average quality of the pups. This is precisely what selection intensity means. The smaller the percentage of pups retained for breeding in each generation, the greater the selection intensity. It is only common sense to have as many young dogs as possible among which to pick and choose, provided other factors are equal; that is, not to overcrowd the accommodation or economize on feeding. Calculating the proportion of breeding animals retained in each generation gives a simple numerical index of the selection intensity. The smaller the intensity, the greater the potential progress towards the ideal.

The simplest form of selection is that of individual merit. That is, selection based upon the appearance of the animal. This is absolutely basic, regardless of whatever other refinements may be employed. There are some characters which are so commonplace that their importance is sometimes overlooked. The first of these is that the dog or bitch should be healthy, as indicated by sound growth, freedom from disease as a pup and lively behaviour. If a male, he should be morphologically normal, display keen interest in the oestrous female and be persistent in coitus. He should have a regular oestrous cycle, mate easily, have satisfactory parturitions, produce normal and healthy offspring in average numbers for the breed, and be able to rear these unaided or with the bare minimum of help. Good health and reproduction are of paramount importance. There is no point in creating a superlative strain only to find that it is held in low esteem because of poor health or reproductive capability.

The most remarkable results are obtained when it is possible to select for a single easily recognized character. These are the cases which produce the nice upward curves, showing the progress of selection for successive generations, to be found in genetic textbooks. Alas, in many situations the practical breeder may find that he or she has to settle for something less. There are numerous reasons for this, of which three may be singled out. The first is numbers. To obtain the maximum amount of progress, large numbers of progeny must be examined for each generation. This is impractical for a slow breeding animal like the dog. However, an effort should be made to breed as many offspring as feasible in the circumstances.

Secondly, the breeder must have a clearly defined goal in front of him, so that the selection can continue generation after generation. It is fatal to chop and change between generations. One must have an objective and be able to stick to it as far as it is possible. Of course, one must exercise judgement. If it is obvious that to adhere to one's original plan will lead nowhere, then it must be abandoned, even if it means returning to square one. Part of the skill of animal breeding is to know when to change.

The third reason is much more fundamental. The breeder is rarely, or never, in a position where he or she can concentrate solely upon one character. Many characters have to be considered together if all-round excellence is to be achieved. There are three methods of tackling the problem and these will be described in turn.

It might seem feasible to concentrate upon one character at a time, moving on to the next when the first has reached a sufficiently high standard. This is the so-called "tandem" method of selection. Unfortunately, the method has several failings. Considerable time is required to obtain worthwhile results, even if only a few important characters are dealt with in turn. It is very difficult to maintain the excellence of the early characters while attention is being given to the later characters. Deterioration usually follows because it is impossible to breed enough offspring to maintain past achievement and to improve yet another character.

The second method is to resolve to retain only those individuals which are of a certain standard for each character. This is the method of "independent culling levels". It has the obvious merit of allowing many characters to be considered simultaneously. It is also easy to apply once a grading scale is instituted, so that each character can be scored objectively. The culling level for any character may be high or low. If these are set too high, so many offspring will be rejected that it may be impossible to find sufficient meritorious young stock to form the next generation. A similar impasse will be reached if a large number of characters are considered.

There are three ways of avoiding the impasse. Only consider the really important characters in the hope that the unimportant will take care of themselves. Consider as many characters as necessary, but lower the culling level for each. Or, vary the culling levels between characters, setting them high for important characters but low for the unimportant. This last procedure does introduce some degree of flexibility into the method. It is the inflexibility which is the primary drawback of the method. The method would exclude an individual which excels in a number of characters simply because it fails in one. Strictly speaking, if the method is to be followed, no exceptions should be permitted, certainly no arbitrary exceptions. If exceptions are to be made, it is wise to have these built into the selection procedure from the beginning.

It is patent that the more well-thought-out the method of selection, the more effective it is likely to be and the greater the overall improvement. One merit of the method of individual culling levels is that it compels the breeder to carefully consider the various characters which contribute towards the ideal dog of his dreams and to observe how these vary from animal to animal. It is surprising how many breeders could not answer this sort of question if challenged. It would seem that a method akin to that of the independent culling level but more flexible in its operation and more embrasive in scope is required. It is possible to devise such a method and this will be discussed in detail. It is to be recommended in preference to either the tandem or independent culling levels.

The third method of selection is known as the "total score" or "selection index". It is a little more complicated to operate than the method of independent culling levels, but this is compensated for by its greater efficiency. The first step is to grade each character upon a numerical scale of merit which should be kept as simple as possible compatible with accuracy. Even a simple grading of five divisions, such as very poor (1), poor (2), average (3), good (4) and very good (5), gives a 5-point scale. Thus a dog poor in one character and good in another would be scored as 2 and 4, respectively, for each character. The scoring is continued over as many characters as desirable. However, an endeavour should be made to score on a finer grading scale than this, such as a 10-point scale which most people are capable of using. This would range from (1) for the poorest expression of the character to (10) for the highest possible grade. Incidentally, the act of representing the quality of each character by a number is known as scoring of that character or simply as "scoring".

One of the advantages of the total score method is that there is no limit to the number of different characters which can be scored. In fact, it is advisable to include as many as practicable, because it is possible that a relatively minor character may turn out to be of importance. In this event, it is useful to be able to examine how the character has been behaving in animals of past generations. It may be that the character has shown a gradual deterioration, in which case it is obvious that more attention must be paid to it in the future. This comment is simply illustrative, for there are many reasons why it may be desirable to check out the variation of a character over successive generations.

The next step is to distinguish between the relative importance of the various characters in a systematic manner. Some characters are more important than others, and these should exert more influence upon the total score than others of less significance. This is achieved by arranging the total number of characters in order of importance, starting with the least important. The relative importance is indicated by assigning a number to each character, giving the least important character(s) the numeral 1. Any character which is thought to be twice as important is given 2, any character which is thought to be thrice as important is given 3, and so on. This procedure of allowing for the importance of the character scores is known as "weighting". It is not necessary for the weights to vary greatly. Thus far, in practice, it has not been necessary to have weighting coefficients greater than 20. That is, many characters were judged to be of equal weight and the most important not more than 20 times that of the least. One reason for not making the coefficients too large is to keep the arithmetic within reasonable bounds, although with a hand calculator this aspect is not a problem.

The grading scale of points should remain unchanged, provided it is properly drawn up in the first instance. However, it may be necessary to change the weighting coefficients from time to time, although it must be cautioned that no changes should be made without good cause. For example, if a character is showing no signs of improvement, it may be necessary to increase its coefficient so that dogs with exceptionally good expression will score relatively higher.

The total score is calculated in the following manner. The dog is scored for each character in turn, awarding points as indicated by the expression of the character. These are then multiplied by the respective weighting coefficients and summed. The sum is the total score for the dog in terms of the characters scored and the weighting coefficients. The selection index is found by dividing the total score by the maximum score (i.e. where each character receives the maximum number of points) and multiplying by 100, a calculation which gives an estimate of the worth of the individual on a scale of 0 and 100. Many people find that the percentage figure conveys more to the mind than the total score, which may run into several figures. It is obvious that high scores or selection indexes will represent excellent animals, clearly suitable for retention for breeding purposes.

The use of the total score method is best explained by a fictitious example. The number of characters chosen will be few, so that the arithmetic is easy to follow. In practice, many more characters will be scored, as the individual is subject to a thorough analysis. Suppose that the following characters are scored: health/condition (H) coat colour (Cc), coat texture (Ct), coat length (Cl), body conformation (Bc), stance (S), head conformation (Hc) and ear carriage (Ec). It may be assumed that the relative importance of these eight characters are in the ratios 10, 5, 4, 4, 3, 2, 1, 1. That is, Hc and Ec are considered to be of equal importance, while S is considered to be twice as important as Hc and Ec, and so on. Note some of the groupings; the head as a whole (Hc plus Ec) is considered to be of equal importance to stance (S). The body as a whole (Bc + Hc + Ec + S) is considered to be a little more important than coat quality as a whole (Ct + Cl). This sort of grouping should be taken into account when the characters are weighted and is an aspect of the skill in formulating the total score equation. Also note that the health of the individual is ranked quite highly, as indicated by a large ratio. This is an important item in the make-up of any pedigree animal for the reasons given earlier.

Having decided upon the appropriate ratio for each of the characters (as above), the total score may be written for computation as:

$$\text{Total score} = 10H + 5Cc + 4Ct + 4Cl + 3Bc + 2S + Hc + Ec.$$

Table 6 illustrates the scores for the above characters for seven hypothetical dogs, using a 10-point scale of excellence. The total score is calculated by fitting the respective score values in the equation. For example, for the dog A, the total score is:

$$\begin{aligned}\text{Total score} &= 10(10) + 5(8) + 4(8) + 4(7) + 3(6) + 2(5) + 8 + 8 \\ &= 100 + 40 + 32 + 28 + 18 + 10 + 8 + 8 \\ &= 244\end{aligned}$$

$$\text{Selection index} = 244/300 = 81.3.$$

The maximum score a dog could attain is 300 for this particular equation, but no animal is able to reach this exacting height. The selection index is found by dividing the observed total score by the maximum possible and multiplying by a hundred. The resulting index expresses the worth of the individual on a percentage scale.

Dog A emerges as the best individual and achieves this by showing good to above average scores for all characters. Dog B is almost as good,

TABLE 6

An Illustration of Calculating the Total Score as an Aid for Selective Breeding

	Points grading for each character								Total	Selection
Dog	H	Cc	Ct	Cl	Bc	S	Hc	Ec	score	index
A	10	8	8	7	6	5	8	8	244	81.3
B	10	10	8	6	5	6	5	4	242	80.7
C	10	6	6	6	6	6	4	2	209	69.7
D	10	7	6	5	6	3	33	3	209	69.7
E	10	10	4	4	5	2	2	2	205	68.3
F	5	9	9	8	6	7	4	4	203	67.7
G	10	5	4	4	5	4	4	4	188	62.7

See text for meaning of the character abbreviations and calculations of total score and selection index.

actually scoring better than A for the important Cc character, but alas being a little inferior in several other characters. This failing is sufficient to relegate the animal to second place. Not that dogs C and D are different animals on the basis of assigned scores, yet they emerge with equal scores. Such ties are not uncommon in practice, signifying that the individuals are of equal status from a selection point of view. Dog E scored high for the important Cc character but quite low for most of the others, and this was sufficient to produce a low score. Thus in spite of its apparent excellence, it is a dubious individual from the viewpoint of selection. The same comment applied to dog F. This is a good dog in respect to several important characters, but fails miserably in the most important of all, viz. health (H). It is assumed that the animal is a bad doer, always appearing somewhat off-colour and thus must be penalized. Had the animal been fully healthy and awarded a score of 10 for H, it would have attained a total score of 253, or a selection index of 84.3, the highest of the table. Dog G is a mediocre specimen for all characters.

The total score is essentially an aid to selection of suitable dogs for future breeding, and its usefulness is dependent upon three factors. Namely, the ability of the breeder to decide those characters which contribute towards the ideal individual, his skill in correctly weighting the characters in order of relative importance and his proficiency to score each character consistently and objectively. This latter aspect is of vital importance, because he or she will be fooling no one but himself or herself in not scoring each character objectively. Giving an individual somewhat better scores than it actually deserves, from sentiment or other reasons, will hinder the breeding programme. Viewing certain animals pessimistically and underscoring will also be a hinderance to the programme. Biases of this nature could be unconscious, and one should be on the alert for the possibility.

A person who is unfamiliar with the total score method may find that some assistance is required to construct the basic equation for the first time. To define the various characters, breaking a composite character into its components or subcharacters and to appropriately weight the various characters. Some experience is necessary before a person can acquire the "feeling" for a soundly constructed total score equation. It may be advisable to draft several provisional equations before a satisfactory one can be found. Do not hesitate to commence with a simple equation initially, adding new characters or subdividing existing characters to get the equation "just right".

It may be that a little experience is required in scoring each character objectively. This is a form of judging of each animal but more analytical in concept. Here again, some assistance could be beneficial for the novice since it is no use in having a vague notion of how a character should appear in a champion dog. One needs to be able to define this as precisely as possible or at least to have a good idea of how the character should appear.

It may be advisable for the novice to seek explanation of the variation of expression of the various characters before embarking upon a serious attempt at scoring. Several trial runs would certainly not be amiss. The facility of quick and accurate scoring only comes from practice. It is a knack which expert judges have to master, even if they are not subjecting the dog to the same amount of critical scrutiny.

It may be wondered if the items making up the total score could be based upon the standard of excellence for the breed. In general, this will not be the case. The standard could serve as a guide in formulating the total score equation, but here the resemblance ends. The total score in most cases will consider more characters than the standard, unless the latter is unusually detailed. The weighting coefficients should not only reflect the relative importance of the characters but also the record of improvement over the past generations. Initially, the coefficients could correspond roughly to the order of importance in either the mind of the breeder or to the points allocation of the standard. However, as the breeding progresses, those characters which do not display any improvement should be more heavily weighted (i.e. the ratio should be increased). The total score should be periodically adjusted. This review of the functioning of the total score is important, but should not be made too frequently since the total score must be given time to show what it can do. As a guide, such changes should not be made at intervals of less than three or four generations.

This does not mean that the total score equation should be regarded as inviolate. The rule is that changes are permitted, but only from good cause, never from an idle whim. Another adjustment which may be necessary is when a character is divided into subcharacters. This could arise as the breeder gains experience in scoring and he or she is forced to conclude that what was thought to be a simple feature is in fact more complex and is composed of independently varying characters which should be scored separately. Adjustments of this nature should be viewed as an extension of the score, and reflects the breeder's increasing proficiency in scoring. They should be made whenever convenient. This means that the old and new total score values will no longer be comparable for the assessment of past and present achievement, but the selection indexes will still be roughly comparable, since the index is independent of the absolute value of the total score.

Those readers who are familiar with the more scientific construction of selection indices may query the arbitrariness of the choice of weighting coefficients. It is possible to derive coefficients which are more directly related to the inherent properties of the observed variation and which should maximize the improvement. Unfortunately, the derivation of these is beyond the level of this book. However, details of their calculations may be found in Lerner (1958), which is an excellent book to consult on most matters of livestock improvement.

Family Selection and Progeny Testing

The simplest form of selection is where all of the pups of one generation (as far as it is practical) are compared to choose the better individuals. This implies that familial relationships are ignored and even pups from mediocre parents may be considered. Should these be considered seriously, or are they merely "flashes in the pan"? On balance, these animals should be considered, particularly if all of the parents are collaterally related. Where the closed stud method of breeding has been adopted, these exceptional individuals must be retained, since their retention is part of the system.

If familial relationships are to be taken into account, there are several methods of doing this. The simplest, and certainly the most direct, is to give preference to the progeny of certain parents; in short, the better individuals of the stud. It may be felt that the progeny of such parents deserve extra consideration. In a sense this is true, and it is a factor which can be kept in mind. However, it is probably not so effective as the method described in the next paragraph.

A superior method is to focus attention on the progeny of different parents, rather than upon the parents themselves. The relative merit of the progeny for each parent may be assessed by their average quality. The total score offers an easy way of doing this by summing the individual scores and dividing by the number of pups. The value obtained is the average score for the progeny. Those parents with the highest progeny average are clearly producing the better offspring. "Family selection" enters the picture when the progenies with the highest average scores are retained and the other progenies are rejected. Individual selection of the better animals within the superior progeny group must still be applied. In fact it will be patent that selection is operating at two levels: between individuals and between parents. The effectiveness of familial selection stems from the fact that the selection can act more drastically upon the underlying genetic constitution of the characters.

In comparing the progeny groups, it is likely that the same parents are involved in different combinations, some of which have produced better progeny than others. This is known as "nicking" in breeder's parlance. The two parents produce exceptional offspring when mated together, but inferior offspring when mated to other partners. Any breeder who finds such fortunate parents should attempt to breed as many offspring as possible from them. These should be given preference in selecting individuals for the next generation. It is possible for the reverse phenomenon to occur, when certain combinations of parents produce only inferior offspring. Admittedly, these decisions should not be made upon a single litter of a few pups, and this aspect places a restriction upon what can be accomplished.

It is possible to systematically exploit differences in the breeding capability of individuals. This can only be attempted with males because bitches cannot produce the numbers of pups to be examined. However, in theory at least, two or more stud males can be usefully compared. The method is to mate the dogs to be compared with the same bitches to produce two rounds of litters. The dog which produces the better average quality progeny will emerge as the animal with superior breeding capability. It is assumed that both will be good dogs phenotypically, but that with the better phenotype need not necessarily be fathering the best pups. This procedure is "progeny testing" in one of its explicit forms, whereby the genetic value of the individual is assessed by the average quality of its progeny.

The above method is used by some breeders to choose a suitable dog to succeed their existing stud dog. The prospective stud is mated to several bitches who have already borne litters by the existing stud. The average of the two groups is compared and if the average of the prospective stud is superior to the existing, this is taken as a point in its favour. Other aspects being equal, the new dog should take over the duties of the old. Sons of stud dogs under the closed stud system may be tried out in this manner in order to check if their breeding capability matches up to their phenotype. In other words, the old stud dog is not retired permanently until the son has proved his worth.

Systems of Mating

There are many systems of mating, but there is little need to consider all of these. The main features can be covered by considering a few basic and general types in the knowledge that the remainder will not be so very different. At the onset, it may be advisable to briefly discuss "random mating" and what is meant by the term. The mating of dogs is carefully regulated (except for mongrels) and there is no suggestion that pedigree dogs behave promiscuously, rather that the owner's decisions are independent of one another in their choice of mates. As far as the spectrum of genes are concerned within a breed, this is random mating. Random mating in a strict sense implies the occasional mating of brother to sister, half-brothers, cousins, etc., yet many breeders may be studiously avoiding these. On the other hand, a few breeders may be exercising inbreeding. Some, in fact, may indulge in it now-and-again, just as others may make the odd wide outcross. On this basis, the notion of random mating within breeds may not be an unreasonable assumption.

Inbreeding, by definition, means the mating together of closely related individuals. In practice, the actual mating in terms of accepted blood relationships should be considered, because this determines the intensity of inbreeding. Some forms of inbreeding must be regarded as close, others as moderate and still others as mild. The intensity of inbreeding can be calculated for most systems of mating, and this is an asset since in some situations it might be desirable to closely inbreed but in others to mildly inbreed.

Inbreeding may be the act of mating related individuals, but what is achieved by it? Summarily, if continued for many generations, it will produce ever-increasing homozygosity in the offspring. That is, the offspring will become more and more genetically homozygous and come to resemble each other in appearance and behaviour. The common ancestry causes many of the same genes to be received by different animals. Moreover, the chances of receiving similar genes increases progressively for each additional generation. The limitation of number of different ancestors is the key factor.

As mentioned earlier, the proportion of homozygous gene pairs in an interbreeding group of animals is known as the "homozygosity". Conversely, the proportion of heterzygous gene pairs in the group is the "heterozygosity". These two terms are the opposite faces of the same coin, and both are in common usage, since some people like to speak of an increase of homozygosity while others of a decrease in heterozygosity. To put the matter rather crudely, the homozygosity denotes the "genetic purity" of the population while the heterozygosity denotes the "genetic impurity". In this area of animal breeding, any freely inbreeding group is referred to as a "population".

The closest form of inbreeding is the continued mating of full brother and sister, otherwise known more briefly as sib mating. Since the inbreeding is intense, easily appreciated by breeders and can be quickly put into practice, the consequences of sib mating have been investigated in depth. The mating system is simple enough. The better animals from the same parents (not necessarily from the same litter) are selected and mated together. The same procedure is followed generation upon generation.

The consequences of continued sib mating upon the heterozygosity is shown by Table 7. The interpretation of the values in the body of the table is as follows. After one generation, the proportion of heterozygosity is 75 per cent of what it was initially (before inbreeding began); after two generations the proportion has fallen to 63 per cent; and so on. After twelve generations the proportion of remaining heterozygosity is about 7 per cent. The successive decreases are a little unsteady at first, but settle down to a constant ratio to one another (the value in a given generation divided by that preceding it). It is convenient to ignore the initial unsteadiness and concentrate upon the constant ratio, since it is characteristic of any system of mating. It is also a convenient measure of the intensity of inbreeding. The lower the ratio, the greater the inbreeding. For sib mating, the ratio, or inbreeding intensity, is 81 per cent. To view the ratio a little differently, this implies a loss of heterozygosity of 19 per cent per generation.

There is one other system of mating which is equivalent to sib mating. This is where the offspring are mated to the younger parent in each generation. A given animal is mated twice, once to its younger parent and again to its offspring. For example, suppose the father was mated to a daughter; then a son would be chosen to mate with his mother and so on *ad infinitum*. The progressive decrease in the proportion of heterozygosity is exactly the same as for sib mating, with the same ratio of decrease of 81 per cent. Although the two systems of mating have the same effect on the decrease of heterozygosity, they should not be substituted for the other if the same ratio of decreased heterozygosity is desired. When the two systems are interchanged, the ratio of decrease is 84 per cent for the generation in which the change occurred. That is, the decrease is retarded to a small extent.

Another easily operated system of mating is where a stud dog is mated to two half-sister bitches who are full sisters of each other. Two groups of progeny and three individuals are involved in each generation. The mating scheme is as follows. A male is mated to two females, producing two litters (or series of litters from the same female, if a wider choice of offspring is desired, defined as a progeny). For either of the two progenies, the best male is selected, whence the other litter contributes the two best females. The rearing of two litters per generation offers a wider choice of offspring than is possible with simple sib mating. The ratio of decrease of heterozygosity is less than that achieved by sib mating, but is still respectable, having the inbreeding intensity ratio of 87 per cent (or a loss of heterozygosity of 13 per cent per generation).

The above three systems of mating may be said to be regular in that the dogs are mated to a set procedure. Most breeding systems have this property, except for the "closed stud" system which does offer a somewhat flexible approach. This is where a few males are mated to a retinue of females. The stud is closed because no new stock is introduced, nor are bitches sent out for mating. The animals of each generation are chosen from among the litters born within the stud. The closed stud may be composed of any number of males and females; usually of a few males and a

TABLE 7

The Expected Decrease of Heterozygosity for Successive Generations of Brother-to-Sister Mating

Generation	Percentage	Generation	Percentage
1	75	7	22
2	63	8	17
3	50	9	14
4	41	10	11
5	33	11	9
6	27	12	7

Females \ Males	1	2	3	4	5	6	8	10
1	0.81							
2	0.85	0.89						
3	0.86	0.91	0.92					
4	0.87	0.92	0.93	0.94				
5	0.87	0.92	0.94	0.95	0.96			
6	0.87	0.92	0.94	0.95	0.96	0.96		
7	0.88	0.93	0.94	0.95	0.96	0.96	0.97	
8	0.88	0.93	0.95	0.96	0.96	0.96	0.97	0.97
9	0.88	0.93	0.95	0.96	0.96	0.97	0.97	0.97
10	0.88	0.93	0.95	0.96	0.96	0.97	0.97	0.98
12	0.88	0.93	0.95	0.96	0.97	0.97	0.97	0.98
15	0.88	0.93	0.95	0.96	0.97	0.97	0.98	0.98
Many	0.89	0.94	0.96	0.97	0.98	0.98	0.99	0.99

FIG. 20. The ratio of decrease of heterozygosis for different numbers of males and females with the closed stud method of breeding.

larger number of females. The rule for selection of breeding animals for the next generation is that (i) the number of males is chosen from among the various progenies, one from each until the required number is reached and (ii) one female from each progeny. The best individuals are chosen in each case.

The closed stud breeding system leads to a decline in heterozygosity because of the limited number of parents within each generation. The intensity of the inbreeding varies according to the number of parents as shown by Fig. 20. The cells in the diagram give the ratio of decrease of heterozygosity according to the number of males and females. It is worthwhile studying the values of the ratio, since a trend is apparent. The heterozygotic decrease is only appreciable if the number of males or females is quite small. From a practical perspective, the number of males is the important item since these are generally much smaller in number compared with the females.

With one male and one female, the closed stud system is identical to sib mating, as a little reflection will confirm. It is only possible to have one progeny, from which it is only possible to choose two individuals, and these are necessarily brother and sister for the ensuing generation. With one male and two females, there are two progenies (same sire but different dams), from which it is possible to have two mating types (male with a full sister and male with a half-sister). This is because each female must contribute one female to ensure the constant number of females per generation. This has an immediate effect on the intensity of inbreeding, as shown by the ratio of 85 per cent. This is a sharp fall compared with the ratio of 81 per cent for sib mating. Nevertheless, the decrease is still appreciable and is accumulative over the generations.

It is interesting to compare the above closed stud system of three individuals with the mating of one male and two half-sisters. The intensity of inbreeding is 81 and 85 per cent, respectively. That is, the closed stud system has the higher rate of inbreeding. This follows because one of the two matings of the system is a sib mating, which is prohibited in the case of one male and two half-sisters. However, one male and three females has the intensity of inbreeding of 86 per cent which is less than the case of one male and two half-sisters. Additional females will further reduce the inbreeding. The decline will continue until it reaches the limit of 89 per cent for one male and very many females (see column one of Fig. 20). Generally speaking, any number of females up to six will give a steady decrease of heterozygosity which should meet the needs of most breeders.

Consultation of Fig. 20 will reveal that the use of two males, even with only a few females, changes the picture sharply. With successive additions of either males or females, the ratio of decrease of heterozygosity rises steadily until the loss is so miniscule that the system of mating can scarcely be regarded as inbreeding in any practical sense. On the other hand, this property is one of the advantages of the closed stud system since the breeder can decide upon any level of inbreeding he or she wishes by regulating the number of males or females.

It must be pointed out with emphasis that when several males and females constitute the closed stud, each must have opportunity to contribute equally to the offspring. If one or more are either favoured or disfavoured, this effectively reduces the number of animals and the inbreeding will be higher than that supposed. This may not always be possible in practice, but every endeavour should be made that each parent contributes equally to the next generation.

The decline of heterozygosity over the generations can be approximated by continued multiplication of the pertinent ratio for any size of closed stud. For example, take a closed stud composed of one male and three females. The decline in heterozygosity after one generation will be 86 per cent, after two generations 86 × 86 = 74, after three generations 86 × 74 = 64, and so on. In this manner, a series of percentages will be derived which are similar to those for sib mating.

In general, the decline will be less rapid than for sib mating, indicating that the intensity of inbreeding is less. The use of the ratio (which is an approximation) will probably overestimate the decline in the early generations, and this point should be borne in mind. However, the values are accurate enough to yield an insight into what is happening to the heterozygosity. In particular, it affords a comparison between closed studs of different compositions.

Should the occasion arise that the ratio of heterozygotic decline need be calculated more exactly than the values given in Fig. 20 or for numbers of males and females outside the scope of the figure, the following formula should be used:

$$\text{ratio} = \tfrac{1}{2}\left[\,1 - 2A + \sqrt{(4A^2 + 1)}\right]$$

where $A = (M + F)/8MF$ in which M = number of males and F = number of females.

The ratio of decrease of the proportion of heterozygosity should not be taken too literally for several reasons. These may be briefly mentioned here, although their impact upon practical breeding is slight. The ratio is quite general in that it is independent of the initial proportion of heterozygosity and of the number of genes involved. It applies particularly to autosomal genes, but less accurately to sex-linked genes. Since sex-linked genes are in a minority, the error is negligible, unless the inbreeding is prolonged, which is probably unlikely. A more likely source of discrepancy between the theoretical and realized decrease of heterozygosity is that the selection of the most healthy and vigorous offspring may select the more heterozygotic individuals. The consequence is that progress towards homozygosity will not be as rapid as one might expect. However, only exceptionally would a large discrepancy occur and, therefore, the ratio may be regarded as a useful index to the intensity of inbreeding which is taking place for any breeding system.

Another useful descriptive index of the intensity of inbreeding is to calculate the number of generations required to reduce the initial proportion of heterozygosity by 50 per cent. The fewer the generations required, the more intense is the inbreeding. Table 8 shows the generations required for sib mating, closed studs of various compositions and cousin matings. Sib mating is obviously the closest and most effective form of inbreeding. The degree of inbreeding for other systems is inversely related to the number of individuals involved for generation. The cousin matings are systems designed to retard inbreeding (see later), and the table reveals that they are most effective in this respect.

TABLE 8

Number of Generations Required to Halve the Initial Proportion of Heterozygosity for Several Systems of Mating

System	No.	System	No.
Sib mating	3	1M, many F	6
1M, 2 sisters	5	2M, 6F	8.8
1M, 3F	4.6	2M, many F	11.4
1M, 5F	5	4 cousins	8.3
1M, 8F	5.5	8 cousins	18.9

M = male(s), F = female(s).

It is not unlikely that it may be impossible to maintain a constant number of breeding individuals per generation. An individual may turn out to be sterile or regrettably die, or some individuals may be of such a high standard that it is decided to breed from them instead of discarding them in the usual manner. Could this have much effect upon the progress to reduce the heterozygosity? If the numbers do not vary too much, the answer is no. On the other hand, should the number vary considerably, then the effect may be significant. The simplest method of tackling the problem is to look up the appropriate ratio within each generation and multiply these together. This will give an estimation of the remaining heterozygosity.

For example, suppose that a closed stud was composed of a single male and the following number of females over seven generations of breeding: 3, 6, 4, 10, 12 and 15. The pertinent ratios are 86, 87, . . . , 88. Multiplying these successively gives an estimate of 39 per cent as the remaining heterozygosity. The smaller numbers of individuals have a proportionately greater effect than the larger. This is shown by the fact that the same reduction of heterozygosity would be achieved by seven generations of one male and a constant number of six females. This is a general rule, the smaller number of breeding individuals will always disproportionally outweigh the larger.

To Inbreed or Not to Inbreed?

The dilemma whether or not to inbreed is not an easy one to resolve. However, one aspect is certain: if a breeder wishes to find a strain of superlative dogs, with its own characteristics and uniformity of offspring, some degree of inbreeding is essential. The reason is that a pure breeding or largely homozygous strain cannot be developed by any other means. Selection alone cannot do it. It can do it spuriously, in that the parents may have similar phenotypes because a certain type has been selected. But what of their progeny? These may be variable, with only an occasional pup tending to resemble the parents. In breeder's jargon, inbreeding is necessary to "fix" those characters which are being so assiduously selected.

There is no doubt that inbreeding is valuable in stabilizing the results of selection. Does this imply that the inbreeding should be deferred until the selective breeding has proved to be successful or should it accompany and perhaps even aid the selection? There is some merit in the idea that inbreeding should be deferred for a few generations, primarily because the most significant advances with selection are usually accomplished during the early stages. Eventually diminishing returns set in and it is more difficult to make further progress. It is often a case of consolidating gains. This is where the inbreeding could commence. This is a simple and straightforward philosophy of breeding. The selection is not relaxed and the inbreeding can commence whenever the breeder considers it most opportune to do so.

However, this is not the whole story and it is worthwhile to examine what inbreeding can do. In practice, there are two systems to consider seriously. These are sib mating and the closed stud system. Sib mating is the most intense form of inbreeding and leads to rapid fixation of characters. Now, this fixation can be a mixed blessing. While it is desirable to fix those characters which produce an outstanding dog, other, less desirable, characters will also be fixed. Furthermore, once fixation has occurred, further changes cannot take place. The characteristics of the kennel are frozen in one mould, so to speak, and additional selection will have little or no effect. On the other hand, of course, if the dogs of the kennel had attained a high standard of perfection, the fixation may not be deemed a disadvantage. More likely, unfortunately, the fixation will have occurred before the full potential of the stock had been reached.

What should be done? It would seem that a compromise is in order. Inbreeding will fix both good and bad characters with fine impartiality. If the selection has been successful, the good features should outweigh the bad and the quality of the strain should reflect this. Suppose, now, that the inbreeding had been of a less intense sort and that fixation had proceeded less rapidly. It should be possible for selection to boost the good points of the stock and eliminate or suppress the bad before fixation occurs. In a nutshell, a balance has to be found between a rate of increase of homozygosity which will bring into being fixation of desirable features yet give sufficient elbow room to eliminate the undesirable.

The closed stud system would seem to foot the bill admirably. The intensity of inbreeding can be controlled by varying the number of animals, especially the number of sires. It is difficult to recommend an optimum size. It would seem that one stud male will be satisfactory in most cases. More than two males may be used (perhaps only temporarily), but the rate of loss of heterozygosity is too small except for very long-term projects.

A typical closed stud could be composed of one male and a variable number of females. In this, the inbreeding may be taken to be moderately-high to moderate, depending upon the number of females. The kingpin is the common sire which tends to hold the stud together. Within each generation, the dogs retained for breeding will be determined by the need for replacement and how strictly the selection is applied. At least one litter must be obtained from each bitch, and selection based upon the total number of pups. It goes without saying that only the best male pup is retained, regardless of the actual mother (unless familial selection is being practised). With the females there is a little more leeway. Strictly, each litter should contribute one female pup—the best naturally.

However, it may be that several very good pups are from one mother, even perhaps from the same mother as the male pup. What should be done in this event? Strictly, if the closed stud system is to be followed, there should be no indecision. The result may be due to chance; if so, the chances that it will recur are slim and one should only keep the best pup. On the other hand, suppose the parents have happily nicked? Then, their offspring should certainly be retained for breeding. There is no guarantee that they will pass on this unique combination of genes, but they will certainly pass on the individual genes, and these may come together again in a future generation. The point to be made is that the closed stud should be used in a flexible manner.

It may be, however, that the breeder is concerned that the mating of the better animals drawn from one litter will be indulging in too close inbreeding. There will, in fact, be a reduction in the number of effective breeders or contributing females for the generation. If the breeder feels strongly opposed to this, he or she can avoid it by adhering to the rules for the closed stud system. On balance, the breeder would be advised to proceed regardless. It is unlikely that the event will recur, for if sibling females are retained, the chances of one producing pups which are superior to those of her sisters will be small. Thus, future choice will revert to that of breeding only from females selected one from each progeny. Although the aim may be to have a set number of females per generation, as a control on the intensity of inbreeding, small variations in the number will not produce great errors; in any event, the decrease in heterozygosity can be plotted in the manner described earlier for variable numbers of females per generation.

If the breeder feels that the intensity of inbreeding is too high, it is always open to him or her to retain one extra male. This would make a sharp difference in the rate of inbreeding; much more than if an extra female is retained. If he or she does this, however, it must be realized that the males must be used at equal frequencies, over all females. To give preference to one male, keeping the other as merely a second string, is to reduce the effective number of animals in the closed stud in a similar manner of not choosing one female from each progeny.

The closed stud should remain closed to outside stock or to matings with unrelated males until such time as further progress within the stud has ceased. It is difficult to give a time scale for this, for visible progress may stop after a few generations or it may continue slowly for many. Sooner or later it will become apparent when an outcross is desirable. Either overall improvement has slackened or a particular fault stubbornly refuses to disappear. As will be discussed, the nature of the outcross must be chosen with care. Ideally, the outcross animal should possess all of the good points of animals of the closed stud or surpass them. If the breeder is after a particular feature in which his or her stock is deficient, the outcross animal should excel in most other points as well.

The introduction of fresh genetic material (genes, as a geneticist would say; "blood", as a breeder would say) into a closed stud is an attempt to bring about an improvement. If the attempt is successful, the breeder will doubtless feel very pleased and perhaps not think too much of what might have gone wrong. The use of an outcross could have two unfortunate consequences. It might disrupt a carefully nurtured genetic constitution which is producing some fine puppies, and it might introduce some unexpected faults. For these reasons it is wise to proceed cautiously. On no account substitute an outsider for the stud dog. It is advisable to make the introduction by sending one (or perhaps two) bitches for mating with the chosen outside stud. Examine the offspring of these most critically before committing them to breed with the rest of the closed stud members. If they fall short of what is regarded as an acceptable standard, reject them and seek another dog with which to make the outcross. All this may seem reasonable on paper, but less so in practice, when personal involvement and expense is taken into account. However, these matters should take second place to the breeding policy for the stud.

Stud Male and Grading Up

The stud male is the most important member of the kennel. Not because of his sex *per se* or because some character may be inherited through the paternal side of the family. No, the reason is numerical. A male can sire many more pups than a bitch can hope to mother. His influence will be seen throughout the kennel solely because of frequent use. It follows, therefore, that the breeder should be exceptionally particular in his choice of the stud dog. Fortunately, it is possible to be more selective of males because fewer are required. In general, this means that the breeder should aim for a degree of excellence in the stud dog at least as high as that of the bitches, but preferably higher. This dictum applies whether the breeder owns a stud or if he is sending his bitches to a dog at public stud.

The mating, often repeatedly over several generations, of inferior animals to those of high quality is known as "grading up". The objective is to raise the level of the inferior animals to that of the superior. Beginners to dog breeding often have to resort to grading-up procedures. Top quality animals will always be scarce and expensive. It is easy to advise a beginner to commence with as high quality stock as he or she can afford, but these may not be available. That is, breeders may not wish to part with their better animals, especially bitches. With dogs, the situation is different; a breeder may already possess a good stud dog and be in a position to offer male pups of high quality—at a price. Therefore, a beginner may be able to acquire a good dog for stud, but only second- or third-rate bitches. These latter should not be despised, provided they are from a good strain. The procedure is to mate not only the bitches to the stud male but also his daughters and grand-daughters, etc., in an attempt to impress his qualities upon the offspring. This is the essence of grading up.

A beginner who does not wish to or cannot purchase a worthwhile stud dog could follow the same procedure of grading up by having the bitch(es) mated to an outstanding sire who is at public stud and sending the daughters and grand-daughters, etc., back to him. Should the chosen stud die or be withdrawn from public stud, the best policy would be to continue the matings to a brother or, better, a son if one is available, depending on important factors of quality. The reason is that it is wise to keep within the same strain, especially if it is renowned for consistency. In general, grading up is the quickest method of improving mediocre animals. Far superior in this respect to "going it alone" by selective breeding, if only mediocre stock is available. The backcrossing should continue until the offspring have reached the level of the superior stock or it is apparent that further improvement cannot be expected.

It may be necessary to warn against a breeding policy which has some aspects of grading up but is not so in a general sense. This is where a mating is made to a superior male and, from the progeny, the best son is chosen and mated with the inferior stock. From this latter progeny, another son is used and so on. A little thought will reveal that the back crossing is in the wrong direction (i.e. to the inferior stock repeatedly, instead of to the superior male repeatedly). Any good features introduced in the initial cross will be progressively weakened. There is only one situation in which this form of backcrossing is feasible, and this is where the desirable feature is produced by a dominant gene. This is unlikely to arise except in a very small number of cases.

Another reason for sending bitches to males at public stud is for the improvement of specific characters. It may be that the dogs of a kennel are of a high quality but are somewhat deficient in a certain feature or features. No amount of selection seems capable of bringing about an improvement. The most probable explanation is an absence of genetic variation for these features. The only means of rectifying matters is an outcross to a stud excelling in these particular features. If the breeders' stock has many good qualities already, extreme care must be taken not to upset these while the new characters are being introduced. The choice of stud must be carefully managed and the subsequent breeding must be handled adroitly. Unlike grading up, where a general and continuing improvement is being sought, this particular mating is a once only operation, to introduce the desirable feature. The next move will depend upon whether the feature shows the desirable change, and whether this can be incorporated into the stock without too much interference or deterioration of the qualities it now possesses. Once the outcross has been made, the offspring should either be inbred or mated with others of the stud, depending upon circumstances.

The subsequent steps do not lend themselves to any worthwhile generalizations. Only the breeder, with full knowledge of his dogs, can decide. Much will depend upon the problem being tackled and the response of the stock to the outcross. If this has been unsuccessful, in that no improvement is evident, a second outcross may be necessary. It would be advisable to choose another stud or even a stud of a different strain.

Whether to use or not to make use of dogs placed at public stud is an open question. There are disadvantages as well as advantages. There is no doubt that the dogs at stud will be excellent representatives of their respective breeds, as witnessed by their wins at show and/or championship status, but this does not necessarily mean that one picked at random is ideal. It may be that the stud may exhibit the same faults as the breeder is trying to correct in his own kennel. A wise person would visit shows to examine the various dogs at first hand. One or more may emerge as the appropriate partner for one or more bitches. Arrangements may then be made for a visit.

Prepotency

The idea of "prepotency" is a breeder's concept. The term is used where a given individual has the remarkable ability of producing offspring bearing a strong resemblance to itself. Males are usually referred to in these terms, partly because they have greater opportunity to reveal the ability than a female and partly as the mystic of a "potent" stud dog. The latter aspect should be regarded as sales propaganda. The idea that one sex is more prepotent, usually the male, because of its sex *per se* is erroneous. The male is an important individual, but this is for his ability to produce greater numbers of offspring, not because of his sex as such.

However, is there a genetic basis to prepotency? Such individuals do arise by chance, and the most likely reason is that they have become homozygous for a group of genes controlling a prominent feature of the breed. The individual will gain quite a reputation and be a useful animal for breeding, especially if it is a male. The interesting aspect is that the more inbred an individual may be, the greater are his chances of being homozygous and of displaying prepotency. This is where an individual from an inbred strain will score over one of heterogeneous origin, even if the latter be superior phenotypically. The prepotency need not be due to dominant genes in the case of an inbred animal, since it will be transmitting a uniform set of genes in every germ cell which a heterozygous individual cannot do. Hence, on average, the offspring may show greater resemblance to the inbred parent. The prepotency may appear less striking but be more influential and probably beneficial in the long term.

Breeding for Intermediate versus Extreme Expression

One of the problems which confront breeders is how to deal simultaneously with characters with differing grades of ideal expression. All breeders have in their mind's eye a mental picture of the ideal dog. Some of the characters which make up this paragon will rely upon an intermediate grade of the total expression, while others will demand an extreme expression. This is a problem, since the two modes of expression require different handling if they are to be achieved and these conflict with each other.

In the case of intermediate expression, the selection will be directed towards eliminating those individuals which deviate most from the ideal, either in a positive or a negative manner. It is a matter of fixing the character around the ideal, and the quickest method of achieving this is to closely inbreed. It is not a question of striving for something not quite obtained, since the ideal phenotype is available. It is a matter of minimizing the variation.

Selection for extreme expression, on the other hand, implies striving for something not yet obtained or is found in only a few outstanding individuals. Extreme expression is produced by groups of polygenes operating in a marked plus or minus direction. Such animals will only be produced by collecting together the appropriate polygenes which are scattered among the breed population at large. This means searching for and breeding from any dog with a phenotype tending in the right direction. To achieve this there must be an opportunity for the polygenes to recombine and engender animals with extreme expression. In other words, there must be maximum variation. Thus, close inbreeding would be inappropriate.

Here resides the conflict. For fixation of intermediate expression, the heterozygosity should be eliminated or at least reduced; for realization of the even more extreme expression, heterozygosity should be encouraged. A compromise must be found, so that the two processes can occur together in the same breeding programme. This usually takes the form of a moderate intensity of inbreeding, coupled with intense selection at all times, in the hope that there will be a steady fixation of intermediate characters but not too quickly so that there can be progress towards extreme expression of other characters.

Since selection of the best dogs is the most important item and should not be relaxed, except in the most extreme circumstances, the variable factor is the amount of inbreeding. There are occasions where intense inbreeding (sib mating or backcrosses to parents) may be necessary, but these are the exception rather than the rule. Selection may operate directly upon the relevant characters if these are few, or indirectly, but nonetheless just as surely by means of the total score when many characters have to be considered. To this is added a moderate degree of inbreeding as provided by the closed stud system of breeding. The inbreeding may be interrupted on occasion by carefully chosen outcrosses to outstanding dogs. The above is a summary of points made in earlier sections.

Like and Unlike Matings

The pairing together of animals which have many characters in common is known as "like-to-like" matings. This is sometimes advocated in the belief that this will lead to fixation of characters. Alas, this sort of mating is ideal for perpetuation of characters but not as a rule for their fixation. This can only be achieved by inbreeding. The combination of like-to-like mating with inbreeding is ideal for the fixation of characters with intermediate expression. The method is a form of selection and, in the immediate context, can be recommended. The problem is that the breeding stock will have other characters which differ in other directions. What of these, for they should not be ignored?

The obverse of mating like-to-like is the mating of unlikes. The rationale for this type of mating is that of compensating for faults in one individual by making sure that they are not present in the other. It is implicit that the mating together of animals with similar faults should be avoided. This disadvantage of a policy of unlike matings is that it is negative. True it tends to produce balanced mating, but at the same time encourages heterozygosity. The faults may disappear among the offspring, but will probably reappear in later generations. This is especially the case if there is "overcompensation". That is, dogs which are exceptionally excellent or poor in certain characters are appropriately matched with the intention of breeding offspring with good to average all-round qualities.

It should be obvious that either of these matings have a certain merit and situations may arise which cannot be resolved without recourse to one or the other. The mating of unlikes should be regarded as a short-term expedient and the breeder should attempt to move on to like-to-like mating as soon as convenient. However, the major problem is that in appraising the breeding stock, it will be found that certain characters call for like-to-like mating and others call for unlike matings and no amount of permuting the individuals for mating will reconcile the conflicting claims. In other words, either policy can only be operated in a pure form with a few characters. This is unsatisfactory, since all (or at least as many as possible) characters should be taken into account simultaneously.

All aspects considered, the total score method of selection is superior to either of the methods described here. In fact, it may be observed that the total score will result in a mixed bag of like and unlike matings, with a trend to a preponderance of like matings over the generations if the selection is successful.

Inbreeding Depression and Hybrid Vigour

The fear is often voiced that inbreeding will bring about a decline in the stamina of the stock. This is the so-called "inbreeding depression" and is a possibility which should not be ignored. People who do so, do so at their peril. The loss of stamina is due to increasing homozygosity of polygenes controlling health and vigour. The effect of any one is small and will pass unnoticed in isolation, but the effect of homozygosity of several, as introduced by the inbreeding, could result in an individual which appears sickly from no obvious cause, lags in growth behind its litter mates, is less lively in play or exercise, is prone to any disease which happens along and is generally below par on any index of vitality one likes to name.

In the ordinary breed population, these slightly deleterious polygenes will be present at heterozygotes and their existence would be unsuspected until exposed by the inbreeding. This is not to suggest that all dogs carry these polygenes, or that inbreeding will inevitably produce weak stock. Quite the contrary as a matter of fact; the inbreeding of inherently healthy normal animals will do no harm. On the other hand, it is wise to have some built-in safeguards. This is the reason for including at least one item for health and to assign to it considerable weight in the total score formula.

The deleterious polygenes may make their presence felt in innumerable ways and in a most insidious fashion. Almost any feature of the normal dog may be affected, slightly at first but potentially more seriously if the warning signs are ignored. There may be a decline in birth weight, followed by poor growth, resulting in an underweight individual. It is common for the reproductive performance to be impaired. This may take the form of a reluctance of the male to take an interest in the opposite sex, in his eagerness or frequency to copulate. Sexual maturity may be delayed in either sex. The female may have irregular heat periods and produce litters of below average size. She may be an indifferent mother. The enhanced proneness to disease may be equally variable and manifest itself in various ways. Perhaps only one feature may be affected, or several to a small degree. It is impossible to forecast how the depression will develop. It usually comes on gradually, affecting some individuals and not others. This aspect means that the depression can be combated by breeding only from the most healthy and vigorous dogs.

Should the inbreeding depression become too severe in spite of efforts to counteract its effects, there is no recourse but to outcross to unrelated stock. If the outcross is wisely undertaken, it ought to be possible to preserve most of the better qualities of the strain, or at least not to lose too many. The first-cross progeny are often remarkably hardy and vigorous. It is not always necessary for one of the parent strains to be fully healthy. Even the first-cross from two somewhat weakly parents (provided they are unrelated) can be remarkably healthy. This phenomenon is so common that it has been called "hybrid vigour" or "heterosis". Random bred animals rarely display heterosis because most of their taken-for-granted hardiness is in fact heterosis of a less obvious sort.

Many strains of dog breeds exhibit mild signs of inbreeding depression as a consequence of their inbred ancestry. This may be tolerated, provided the effects are not too severe, because to outcross would destroy their hard-won excellence or their pedigree status. Perhaps the last reason should not be taken into account when the stamina of the dogs is at stake, but it often is. It is possible that outcrosses may fail to restore the stamina, indicating that the depression is widespread and not confined to a few strains.

Breeders of rare breeds of dogs of less than a few score of individuals can suddenly find themselves in the above situation, because after a few generations most of the individuals become interrelated. Inadvertently, the deleterious polygenes have spread throughout the breed. Every endeavour should be made to breed only from the most vigorous animals and to avoid inbreeding in order to prevent a worsening of the situation.

Measurement of Inbreeding

It is relatively easy to measure inbreeding for regular systems of mating. That is, for a set pattern of mating (e.g. sib mating) or for a constant number of breeding animals per generation (e.g. closed stud). The easiest method is to make use of the ratio of the decrease of heterozygosis. Thus, for a closed stud of one male and three females, the decrease for one generation is 0.86, for two generations $0.86 \times 0.86 = 0.74$, for three generations $0.86 \times 0.74 = 0.64$ and so on. To convert this to the increase of homozygosity, subtract the final figure from unity. For example, after three generations the increase of homozygosity is $1 - 0.64 = 0.36$. This is not an exact value, but a reasonable approximation. In the present context, 0.36 is a measure of the amount of inbreeding the dogs have undergone. It is usual to denote the amount as a percentage by multiplying by 100, namely $0.36 \times 100 =$ 36 per cent.

This is all very well for regular systems of mating, but what of irregular systems? A dog may have a pedigree in which one or more ancestors occur

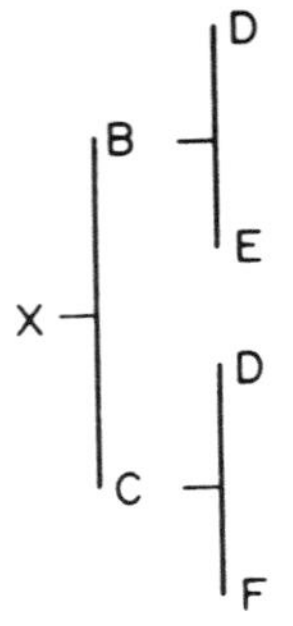

FIG. 21. Pedigree of a half-brother to sister mating. $F = 0.125$.

more than once, indicating past inbreeding. It is possible to estimate the amount of inbreeding in these cases, but a very different method must be used. The essence of the method is the tracing of ancestral paths from the ancestor or ancestors to the individual whose inbreeding is being measured. The method is a measure of the probability that a gene will be handed on from a remote ancestor. Since what happens to one gene may be regarded as typical for all genes, the probability is a measure of the contribution of the ancestor to the genetic endowment of the individual.

The probability that a parent will hand on a gene is 0.5 (i.e. a parent of genotype *Aa* will pass on either *A* or *a* in half of its germ cells). For a grandparent the probability is $0.5 \times 0.5 = 0.25$, the rule being for each additional generation the probabiiity is multiplied by 0.5. To qualify as inbred, an individual X must have a common ancestor on both the sire's and dam's side of the pedigree. Hence, the inbreeding is measured by calculating the probability of an individual receiving a contribution of genes from a common ancestor, say A. This is done by counting the number of generations from A to X on the sire's side and back to A on the dam's side of the pedigree, say n in all. That is, a circular path leading from the sire's A to X back to the dam's A, with a variable number of generations n in between. The measurement of inbreeding is denoted as F, the coefficient of inbreeding for a given individual.

As an example of the method consider the individual X of Fig. 21, where letters indicate different animals in the pedigree. Inspection of the ancestry reveals that the dog D occurs on both sides of the pedigree. The mating is that of a half-brother and sister, D being the common father of both parents. Hence the circular path connecting X to D is BXC. Thus, three generations are involved and $n = 3$. Therefore the amount of inbreeding is $F = (0.5)^n = (0.5)^3 = 0.125$. Note that the path commences at D on the sire's side of the pedigree and ends at D on the dam's side, but D itself is not included. This

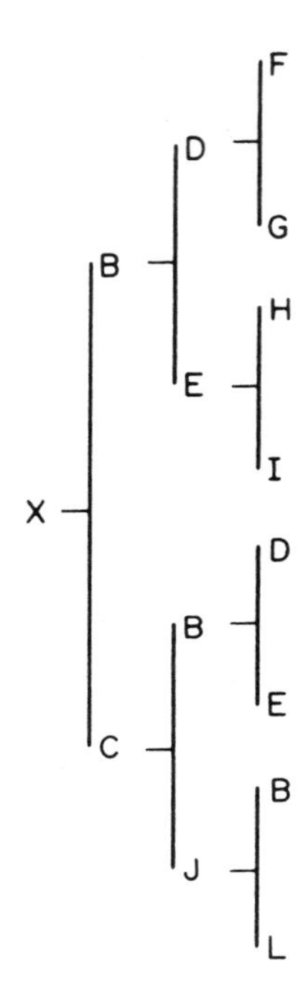

FIG. 22. Pedigree of successive back-cross matings (line-breeding) to a male B. $F = 0.375$.

is the first rule for tracing a path: the common ancestor of the individual X is not counted.

Now examine the apparently more complicated pedigree of Fig. 22. This represents an individual X which is the outcome of two consecutive back-cross matings to the sire B. There are two acts of inbreeding in the ancestry. The first is the mating of the daughter C to B, giving the path XC; the second is the mating of daughter J to B, to give the path XCJ. The amount of inbreeding is the sum of these two paths:

$$F = \text{XC} + \text{XCJ} = (0.5)^2 + (0.5)^3 = 0.375$$

Several items should be noted in the calculations. Common ancestors are not counted, hence the path commences directly with X on the sire's side of the pedigree. The number of generations may differ on each side of the pedigree according to the position occupied by the common ancester. Finally, there may be more than one path to the common ancestor, depending upon the amount of inbreeding that has occurred. This is the second rule for tracing paths: always search out all possible paths in a pedigree.

It may be wondered if individuals D and E contribute towards the inbreeding, since these occur in both the sire's and the dam's side of the pedigree. However, this is not the case. This may be stated more formally

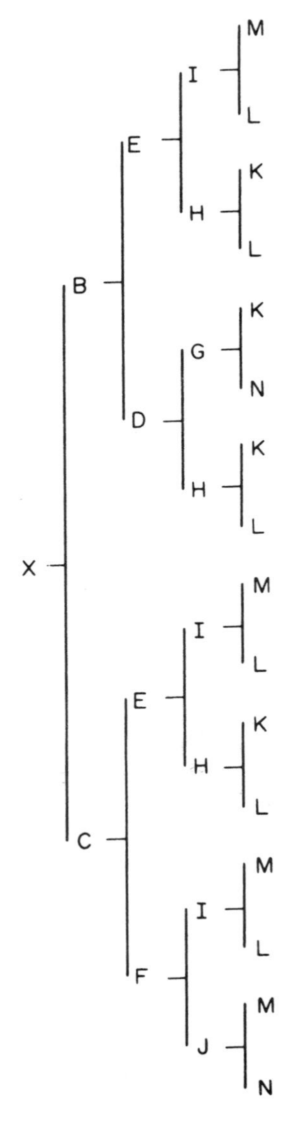

FIG. 23. Pedigree with irregular inbreeding. $F = 0.25$.

by the third rule in tracing paths: no individual must be counted more than once in a path. It might be thought that D or E could contribute towards the inbreeding by a path BXCB, but this path is invalid because B is included twice. The contributions of D and E are in fact covered by that of B.

It is instructive to consider the pedigree of Fig. 22 if the second occurrence of B is replaced by a brother, represented as M, and the third occurrence by an unrelated individual N. This means that B is not contributing towards the inbreeding of X, but that D and E will do so. The path for D is BXCM = $(0.5)^4$ = 0.0625 and the path for E is BXCM = $(0.5)^4$ = 0.0625. Two different ancestors are involved and the total inbreeding F is the sum of the two paths, namely 0.125. This example shows that it is possible to have more than one common ancestor in a pedigree and each one will make a contribution to the inbreeding coefficient F.

A more complex pedigree may now be analysed, such as may occur in practice as a result of haphazard inbreeding over many generations. Such a pedigree could be similar to that of Fig. 23. The working should be done systematically so as to avoid errors. The first step is to list the number of common ancestors. These are the seven individuals: E, H, I, K, L, M and N. The next step is to trace the total number of different paths connecting the same common ancestor on both the sire's and dam's side of the pedigree. The last step is to check that these are perfectly valid. For instance, individual L occurs frequently in the fourth generation and the total number of paths for L may be systematically written as:

IEBXCEI	HEBXCEI	HDBXCEI*
IEBXCEH	HEBXCEH	HDBXCEH
IEBXCHI	HEBXCFI*	HDBXCFI*

However, only three are valid (those with asterisks), to conform with the rule that no individual shall be included more than once in a path. The paths for all other common ancestors must be similarly examined. The final result is the nine independent paths as shown by Table 9. This pedigree reveals that several common ancestors may contribute towards the total inbreeding of an individual. While most common ancestors will have only one path, it is not uncommon for some to have more than one, as in the present case of L.

The calculations are not quite completed; when a number of possible paths have to be rejected, it often means that some of the ancestors are inbred themselves. This will increase the inbreeding of X and must be allowed for in the calculations. This is done by examination of the pedigree and by noting which ancestors are recurring in the invalid paths. In the example, this is individual E. Scrutiny of the pedigree reveals that E is the offspring of a half-brother to sister mating. the inbreeding of E must be

TABLE 9

Calculation of the Inbreeding Coefficient of X from the Pedigree of Fig. 23

Ancestor	Path	Inbreeding
E	BXC(1.125)	$(0.5)^3(1.125) = 0.140625$
H	DBXCE	$(0.5)^5 = 0.03125$
I	CBXCF	$(0.5)^5 = 0.03125$
K	GDBXCEH	$(0.5)^7 = 0.0078125$
L	HEBXCFI	$(0.5)^7 = 0.0078125$
L	HDBXCEI	$(0.5)^7 = 0.0078125$
L	HDBXCFI	$(0.5)^7 = 0.0078125$
M	IEBXCFJ	$(0.5)^7 = 0.0078125$
N	GDBXCFJ	$(0.5)^7 = \underline{0.0078125}$
		$F = 0.25$

calculated as a separate item and is due to the path IEH, giving $F_E = (0.5)^3 = 0.125$. Now, the contribution of E to the inbreeding of X must contain an extra term to allow for the inbreeding of E. This is done as follows:

$$\text{BXC} + \text{BXC}\ (0.125) = \text{BXC}\ (1 + 0.125)$$

as shown by the first row in the table. BXC is the ordinary path for E, while BXC (0.125) is the addition term to account for the inbreeding of E.

The above procedures for the inbreeding coefficient of an individual X are summarized in the following formula:

$$F_X = \Sigma\ (0.5)^n\ (1 + F_A)$$

where F_A is the inbreeding coefficient of the common ancestor A. A is a general symbol to represent any one of a number of ancestors which may contribute towards the inbreeding of X. if the common ancestors are not inbred, $F_A = 0$ and the formula is:

$$F_X = \Sigma\ (0.5)^n.$$

The symbol Σ means summation of all possible paths as explained above.

Application of the formula for the pedigree of Fig. 23 gives the tabulation of Table 9. Summing the entries gives a value of F of 0.25 or, as a percentage, a coefficient of 25 per cent. Despite the large number of common ancestors, the inbreeding is not particularly great. The reason for this is that the majority of contributing paths extend back to remote ancestors. In general, the more distant the ancestor, the less will be its contribution to the inbreeding.

The coefficient F varies from zero (no inbreeding) to unity (completely inbred). It measures the increase in the proportion of homozygosity. The proportion of heterozygosity will by $1 - F$. Therefore, the individual X of the above example shows a proportion of heterozygosity of $1 - 0.25 = 0.75$. This represents the proportionate decrease relative to that present before inbreeding began. Many dogs will be found to be inbred to some extent, but coefficients less than 0.5 are not likely to be harmful. On the other hand, dogs with coefficients over 0.5 must be regarded as becoming moderate to highly inbred and should be carefully watched.

However, it is the rate of inbreeding (amount per generation) which is of importance in judging the significance of the inbreeding. The rate may be found by taking the k root of $1 - F$, where k is the number of generations in the pedigree. In the above example, X is the product of four generations of inbreeding, hence $k = 4$ and $\sqrt[4]{(0.75)} = 0.9306$. Comparing this value with ratios of decrease of heterozygosity in Fig. 20 reveals that it is almost identical with a closed stud composed of three males and four females per generation. Therefore, the rate of inbreeding behind X is not excessive.

Although the calculations may seem formidable, they are not necessarily so in practice if the arithmetic is carried out methodically. The F coeffi cient is the best measure of inbreeding that can be devised, and is employed in many fields of genetics. In dog breeding, the calculation of F can be used to monitor how much inbreeding is occurring in a blood line or breed. It may also be used to check on the amount of inbreeding if a certain stud dog is used for breeding. Many of the top blood lines of breeds are inbred and are interrelated to some degree. It may be prudent to check on the amount of inbreeding which may occur should a projected mating be put into effect. The amount of inbreeding is one of many factors which a careful breeder should bear in mind in planning matings.

When a tabulation is made in the manner of Table 9, the relative contribution of any ancestor can be assessed. In this example, the largest contribution is made by E. This follows for two reasons, (i) the individual E is only two generations removed from X, and (ii) E is inbred itself. H and I are the next largest contributions, followed by L, with negligible contributions from M or N. If E is an outstanding dog, X might be considered to be a likely breeding animal because of the contribution of E. However, the inbreeding coefficient F is low, hence the example is not a good one in this respect. Had F been as large as 0.75, for instance, the contribution of E could be of some significance in evaluating the worth of X.

Avoidance of Inbreeding

On occasion it may be necessary to maintain a small group of dogs with the minimum loss of heterozygosity. Such a situation could be where a stud has to attain a high level of perfection but signs of inbreeding depression

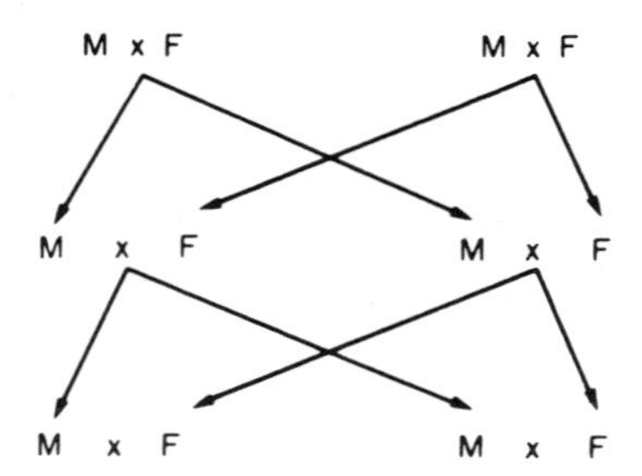

FIG. 24. Diagram of matings for a breeding group of four cousins. Two males (M) and two females (F) are required in each generation. These must be paired as shown by the lines of descent. The pairings repeat for each generation. It is permissible for the sexes to be interchanged.

has been detected. An outcross would be the obvious solution, but this has been judged undesirable, either on sentimental grounds (the breeder may not wish to break up an acclaimed strain) or on the unavailability of unrelated animals of comparable quality. Whatever the reason, other solutions were sought. One would be to increase the size of the group, even if this means a lowering of the usual standard of excellence for stock. At least, a larger number of offspring would become available from which to choose the most vigorous pups. The position may be aggravated if the accommodation only allowed the keeping of a small number of breeding individuals.

A compromise would be to adopt a system of mating for which the decline of heterozygosity is minimal. Several of these may be proposed, among the most efficient being cousin matings. To illustrate the principle, two of the simplest may be described involving only four and eight dogs. The groups consist of equal numbers of males (M) and females (F), which are

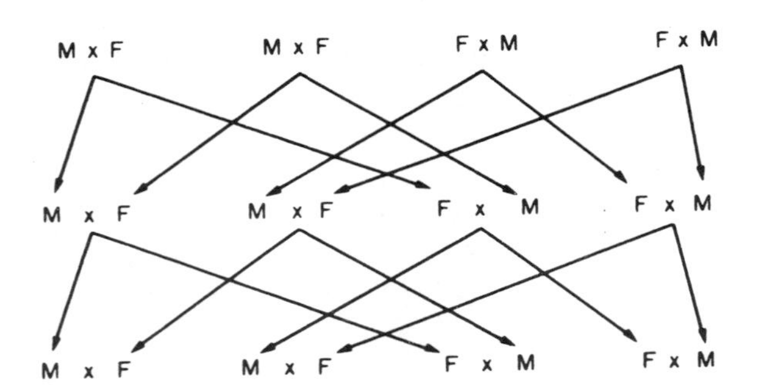

FIG. 25. Diagram of matings for a group of eight cousins. Four males (M) and four females (F) are required in each generation. These must be paired as shown by the lines of descent. The pairings repeat for each generation.

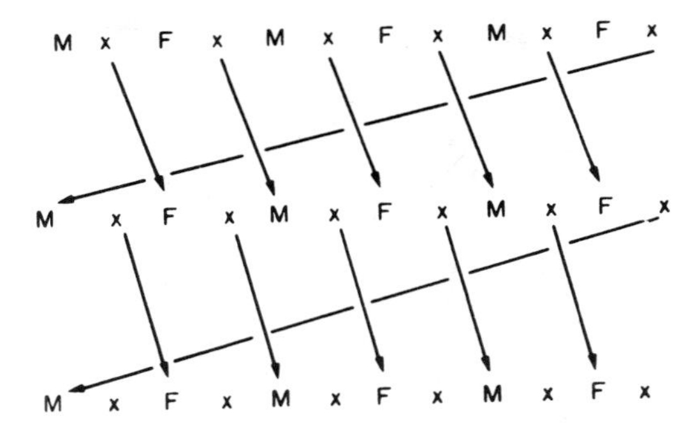

FIG. 26. Diagram of circular matings for a group of six dogs. Three males (M) and three females (F) are required in each generation. These must be paired as shown by the lines of descent. The pairings repeat for each generation.

mated as shown by Figs. 24 and 25. These systems are two of a series known as the maximum avoidance system.

In the first, two males and two females are chosen from the two progenies produced in each generation. The choice of sex for a particular mating is not important, except that a pup's mate must obviously be of the opposite sex and chosen from the other progeny. Figure 24 shows that the matings repeat for each generation in a systematic manner. The ratio of decrease of heterozygosity is 92 per cent per generation, a ratio which cannot be bettered by another system of four dogs.

The second system requires four males and four females per generation. The arrangement of matings is more complicated and is best understood by reference to Fig. 25, although these are merely an extension of those for the previous system. Each of the matings must provide two pups for the next generation, a male and a female. The lines of descent indicate how the eight pups must be mated for the next round of progenies. The whole process is then repeated. The ratio of decrease of heterozygosity is 97 per cent, again larger than other systems for the same number of individuals per generation.

A third system of mating which is simpler to operate than the above is known as circular mating (Fig. 26). Similar to the others, equal numbers of males and females must be chosen for each generation. The easiest way of visualising the mating scheme is to line up the animals in a row, alternating the sexes, male, female, male.... Each animal is mated twice, once to the animal on the left and again to the animal on the right, producing two distinct progenies (litters with different parents). The dog on the extreme right is mated to that on the extreme left as one of its matings. This last mating completes the circle of matings. Each progeny must contribute one pup to the next generation. The sexes must alternate, although the first pup chosen may be of either sex. It is imperative that the correct sequence of mating be followed in each generation.

Compared with the maximum avoidance system, the circular system is a little more flexible. The former can only be used for groups of dogs which increase in powers of two, namely for groups of 4, 8, 16 etc., whereas the latter can be used for any even number of dogs, namely 2,4,6, etc. The illustration in Fig. 26 is for a kennel of six breeding individuals, deliberately chosen to be mid-way between two maximum avoidance systems. The ratio of decrease of heterozygosity is 96 per cent. The circular system compares with the maximum avoidance in the long term, but not in the short term (say less than 50 generations), hence the maximum avoidance system should always be chosen if there is a choice.

In practice, for all systems it is advisable to draught a sketch-plan of the matings, showing the lines of descent and entering the names of the dogs in the appropriate places (replacing the M and F symbols of the figures). The sketch-plan should be completed for the first two generations so that the matings are clearly defined. Subsequently, it is only necessary to add a row of names for the next generation. It is easy to commit errors if an attempt is made to operate the system in the mind or, less systematically, on odd pieces of paper.

Once inbreeding depression has set in, sooner or later an outcross will be necessary to revitalize the stud. The various systems described above should be regarded as a holding operation until suitable animals make an appearance to which an outcross can be made.

One of the above systems could be adopted by breeders of the rarer breeds who are concerned that inbreeding depression could be appearing in their breed. It is difficult to say how small in numbers a breed must be before danger threatens. However, if one accepts the standard adopted by geneticists for the maintenance of laboratory strains of animals, the loss of heterozygosity should not be less than 1 per cent per generation. This implies a breeding group of about 12 to 14 males mated to a large number of females. This means, of course, 12 to 14 actively used stud males, not a situation where a few are used extensively and the remainder only occasionally. The real dilemma here is that an outcross is impossible if the integrity of the breed is to be safeguarded.

A few clubs have attempted to halt any tendency to inbreed by imposing restrictions on the mating of near relatives. The usual stipulation is that no common ancestors should occur in the pedigree of dog X for at least three or four generations. Consider first the implication of the three-generation interval, as shown by the sample pedigrees of Figs. 27 and 28. In the first, the individual X has a common ancestor A in the great-grandparental generation. In the second, X has been mated to a dog A which featured in the G-G-G- parental generation. Both of these pedigrees conform to the three-generation rule. How much inbreeding is occurring? The inbreeding

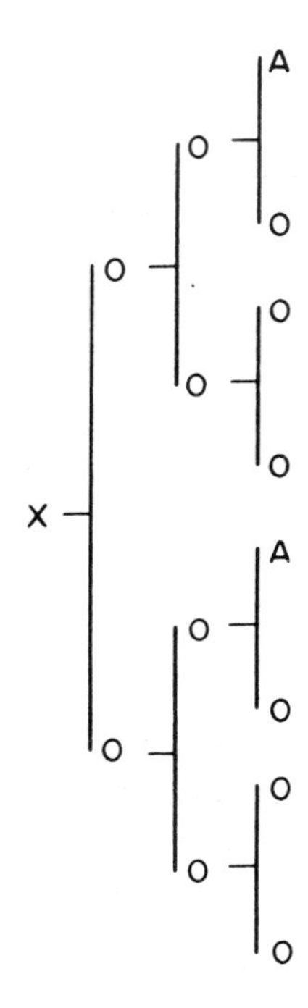

FIG. 27. Pedigree in which the same ancestor does not occur until three generations have elapsed. The individual X has a common great-grandparent A. $F = 0.03125$.

coefficient F for X in the two pedigrees is 0.03125 and 0.0625, respectively. The rate of inbreeding per generation is 1.05 and 1.6 per cent, respectively. These values are reasonably close to the 1 per cent loss of heterozygosity per generation recommended above.

If the four-generation interval before a dog can occur twice in a pedigree is adopted, the same situation could arise as shown in the figure, except that one additional generation is involved. The inbreeding coefficient F for the two pedigrees will now be 0.008 and 0.031, respectively. The rate of inbreeding per generation is 0.2 and 0.6 per cent, respectively. These are smaller rates than those for the three-generation interval and shows the effect of imposing an extra generation. These restrictions force breeders to seek the services of a number of different stud dogs, perhaps several more than they would have done in the ordinary course of breeding. The number of different stud dogs in the three-generation scheme for the two cited pedigrees is 6 and 7, respectively. This is the absolute minimum number, for ideally, sufficient stud dogs should be available for breeders to have a choice as regards suitability in other respects. In other words, the quest to avoid inbreeding should not interfere with the necessity of only employing

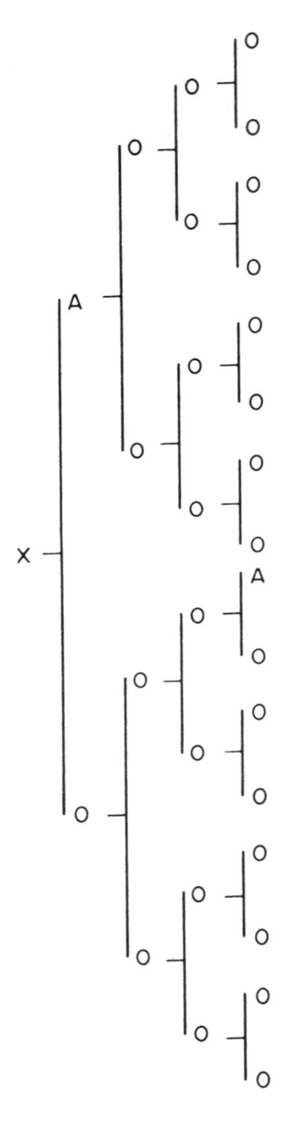

FIG. 28. Another pedigree in which the same ancestor does not occur until three generations have elapsed. Here a back-cross has been made to a past ancestor. $F = 0.0625$.

the best dogs for breeding. The use of inferior animals simply to avoid inbreeding may also be damaging to the breed.

There is little information on the effects of inbreeding on the mortality and vitality of pups or adult dogs. However, the data of Rehfield (1979) for

TABLE 10

Percentage of Newborn Pups Surviving Birth for a Large Colony of Beagles

Inbreeding coefficient F	Equivalency of sib matings	No. of pups examined	Per cent mortality
0	0	489	29
0.25	1	636	25
0.375	2	49	26
0.5	3	76	37
0.594	4	307	50
0.672	5	94	45
0.734	6	17	76
0.785	7	22	73

a Beagle colony is of interest (Table 10). Figures are given for the mortality of pups from any cause according to the amount of inbreeding they have undergone. For comparison, the inbreeding is scaled for the equivalent number of brother—sister matings required to give the same amount of inbreeding. No increase in mortality was observed until $F = 0.594$, when the mortality was just about doubled. These figures are for one colony, and similar data for other colonies or breeds may not be the same. It is not known if these figures can be held to be representative of an aspect of inbreeding. In another colony, the inbreeding may not affect mortality at birth, but may lower the general vitality or reduce body size with minimal effect on health.

The Meaning of Strain

Many people at some time may have felt that they would like to have a strain of dog of their own. This is a natural feeling and prompts the question "What is a strain?" One can, of course, assemble a few dogs and speak of a strain. This attitude is encouraged by the registration of names of studs, but there is much more to the term than that. The registration of stud names has the advantage that it ensures that a stud will become widely known if the dogs from it become regular winners. Almost certainly there will be a demand for pups and there may be requests for the service of the stud dog. A certain amount of inbreeding may be occurring and presently it will become legitimate to speak of a blood line. People will talk of So-and-So's dogs. Eventually the term strain will be used, probably in a general sense. However, the breeder will know that he or she has "arrived" when knowledgeable people remark, "that's So-and-So's strain, I'd know the dogs anywhere."

The term strain should not in fact be lightly used in animal breeding. The term is often based in being applied to a group of meritorious animals which do not breed true. Strictly, the term should be reserved for a kennel of relatively true breeding dogs possessing unique features which make them stand out from their fellows. This sort of reputation is not achieved overnight, and is the worthy culmination of years of unremitting selective breeding. It would also have to be accompanied by a certain amount of inbreeding to ensure that the strain is true breeding. This aspect may be forced upon the breeder because if the animals are that outstanding unrelated dogs of equal quality may be few and far between, and as likely as not to be in the hands of keen rivals.

The term blood line is frequently bandied about rather loosely in animal breeding, usually to indicate relationship rather than anything else. In its most precise sense, blood line means an ancestry tracing back to an exceptional animal or group of animals. This is direct lineage or descent. Collateral or indirect descent becomes of pertinence when it becomes desirable to show that two animals are related via a remote ancestor common to both. Blood lines are often held in high regard by breeders, particularly when notable individuals feature in the pedigree.

Unfortunately, the importance of blood relationship can be easily overstressed. Outstanding individuals can produce mediocre offspring and all pedigrees should be examined with a critical eye. How far back do the exceptional individuals occur, or are they only on one side of the pedigree? The first possibility could be indicative of a slow deterioration of the stock and the second that some grading up has been in progress. In general, high sounding names should not count for much unless one is familiar with the dogs and can assess their worth in relation to the breeder's own stock.

Occurrence of Anomalies

One of the objections which is often raised against inbreeding is that it produces anomalies. This is an understandable objection and deserves to be answered. However, the possibility is sometimes overemphasized. In the first place, inbreeding cannot of itself induce anomalies. What it can do is to bring to the surface any anomaly which may be latent in the stock. Random mating or outcross mating tends to keep recessive mutants covered up except for the odd occurrence here and there. It is even possible for a mutant to be introduced into a kennel, persist for a few generations and be bred out, without anyone being aware of what is happening.

With inbreeding, the situation is different. One of the properties of inbreeding is that it brings out both the good and bad points in the stock. The bad points may include an occasional anomaly. This is the connection between inbreeding and the occurrence of anomalies. The attitude towards anomalies is ambivalent. There are those who would rather not know of any anomaly which may be present, and there are those who, while not feeling very pleased, would rather know and deliberately eliminate it. Some methods of achieving this are discussed in a subsequent chapter. In one respect, inbreeding can be viewed as a "purifying" operation, for if it is instrumental in revealing an anomaly, it also leads to its rapid elimination.

The chance occurrence of an anomaly is not the same as inbreeding depression discussed in an earlier section. It is a related phenomenon, admittedly, and a more vexing problem. It may be stated that the occurrence of an anomaly is rare and certainly not to be anticipated. There are two main reasons why one hears so much of genetic anomalies at present. The first is the considerable interest taken in such conditions by veterinarians and the awareness of breeders to be on the alert for them. The second is the vast numbers of dogs bred all over the world. Although their occurrence at any place or time is extremely low, the enormous numbers of dogs examined ensures that several new anomalies are discovered each year.

A Summing Up

The topics of this chapter are so fundamental to animal breeding that it is desirable to summarize them to round off the discussion. The effects of selection and inbreeding may be briefly contrasted as below:

Selection	Inbreeding
Perpetuation of certain genes	Fixation of all genes
Small decrease of heterozygosity	Steady decrease of heterozygosity
Increasing phenotypic similarity	Increasing genotypic similarity

A simple viewpoint would hold that the consequences of selection and inbreeding are complementary. All that the breeder has to do is to select the right animals and inbreed closely, this leading automatically to fixation of the advantageous genes and the breeding of first quality animals. Unfortunately, it is very difficult to concentrate all of the desirable genes in a few animals so that they can be fixed in the offspring. Additionally, close inbreeding produces a situation in which all of the genetic variability is used up and further progress is impossible. This may be tolerable if the dogs are superlative specimens, but intolerable if the dogs are less than perfect.

A compromise must be found in the form of moderate to mild inbreeding, rather than close. There cannot be any compromise in the matter of selection. This must be as stringent as possible, after making allowance for the quality of the available stock. The principle is for steady fixation of genes, but not so rapid that both good and bad features are fixed before the breeder has had time to weed out the bad. The concept is that of selection holding the genes in the population until the breeding has had opportunity to fix them permanently. In the meantime, some variation is to be expected and countered by the persistent selection.

It will be appreciated that the above recommendation is a general compromise. In certain situations, selection with the minimum or absence of inbreeding may be the best procedure, or selection with intense inbreeding may be most fitting. The degree of inbreeding is often the crucial aspect of any programme of breeding, but this often cannot be decided without reference to the nature of the characters(s) which the breeder is seeking to control. There are no quick and easy solutions, just a lifetime of careful breeding to produce healthy dogs, full of vigour and of handsome appearance, which are a credit to the breeder and no disgrace to the show ring.

6.

Colour and Coat Variation

BEFORE embarking upon an exposition of the genetics of breeds, it is desirable to outline what is reasonably known about the genes and their primary effects. Several of these have been introduced in an earlier chapter to illustrate various basic principles of heredity. However, here the effects will be discussed more systematically and at greater depth. In the main, conventional symbols have been used to denote the various mutants, except for small changes to bring those for the dog in line with standard mammalian genetic nomenclature. There is no justification for the dog to be the odd-man-out in the matter of symbolism, for the species could suffer if the symbols are too esoteric for the non-canine reader.

There have been pitifully few controlled breeding experiments with dogs, probably because of expense and time involved. True, the results of a number of crosses have been reported, even if these are derived from casual, serendipitous observations. Indeed the bulk of the information on canine genetics have been derived from careful analysis of stud book records. These have been most helpful and probably will continue to be so in the future. Anyone who has access to records of this nature are urged to examine them with an eye for new or confirmatory information. This request applies especially to those breeds which have not as yet been studied in detail.

Unfortunately, provision must be made for various types of error. Different people may have different ideas of naming the colours. These show up in the data as apparently spurious results from certain matings. These are acceptable under the circumstances, provided they are not too numerous. Should the exceptions be numerous, on the other hand, the chances are that the hypothesis behind the analysis is incorrect. Individual datum would have to be carefully examined under these circumstances for the source of error, either for false assumptions, mistakes of identification, errors in recording or even incorrect collection of data.

Agouti Alleles

Several of the most common colour phenotypes are due to a series of alleles known as the agouti. The term agouti is taken from a small rodent of this name which has a brown-grey coat, highly adapted for concealment from predators. Among the canids, the corresponding colour is wolf-grey, again adapted for concealment, but this time for stalking prey. The designation of agouti has been carried over to canine genetics from rodent genetics where this type of mutation has been painstakingly studied. This has revealed the type of allele to be expected to arise from the basic agouti gene. In the event, a whole series have been discovered.

The agouti alleles may be represented as follows:

Designation	Symbol
Solid black	A^s
Dominant yellow	A^y
Agouti	A
Saddle	a^{sa}
Black and tan	a^t
Non-agouti	a

The normal gene A is responsible for the wolf-grey colour of wild canids and presumably for the majority, if not all, of dog breeds with wolf-like grey phenotypes, particularly breeds such as the Jamthund, grey Norwegian Elkhound, grey Siberian Husky and others of the far north. The grey German Shepherd dog is probably AA. The gene has given rise to five mutant alleles, two dominant and three recessive to it.

The top dominant A^s is responsible for the solid black variety found in many breeds. The colour varies from coal black to a brownish black, the latter due to a variable number of vestigial agouti-like hairs. Mongrel black dogs frequently have a brownish cast to the coat. Several explanations may be offered for the presence of these agouti-like hairs. The simplest is that the A^s gene is incapable of inducing a solid black coat by itself. It has required subsequent selection by breeders of modifying polygenes to produce a solid black phenotype.

Another explanation is that A^s is completely dominant, so that A^sA^s is solid black while, for example, A^sA^y or A^sA may be brownish black. The agouti-like hairs may be responsible for the "peculiar reddish tinge" noted by Little (1957) which he ascribes to the incomplete dominance of A^s to A^y and a^t. Little has argued that A^sA^y individuals show a reddish undertone, chiefly on the head, neck, sides of the body and legs, but A^sa^t individuals show the reddish tinge only on those areas which are tan coloured in a^ta^t animals. The incomplete dominance of A^s is feasible, but it should be noted that no firm statistical evidence exists for it.

The dominant yellow allele, A^y, is responsible for one of the yellow or red phenotypes of the dog. The A^y yellow can be a clear yellow, but is often found to have a variable number of black-tipped hairs on the head, shoulders and along the spine, including the tail. When the dark hairs are abundant, the sable is produced. For this reason, the allele has also been termed sable or sable-yellow. The all-yellow form has been termed golden sable or clear sable in this terminology. The number of black hairs or degree of sable coloration is variable and due to polygenes which are inherited independently of A^y. These polygenes are known collectively as umbrous, meaning a darkening.

Little (1957) has suggested that sable is due to the heterozygous genotype $A^y a^t$. This would imply the assortment of yellow:sable:black and tan in a 1:2:1 ratio for matings of sable to sable. Sable matings rarely, if at all, produce ratios of this nature. Furthermore, elsewhere in his book Little states that black and tan pups have been repeatedly bred from clear yellow Dachshund parents. Little dismisses these as exceptions to the rule. Unfortunately for the idea of incomplete dominance, these Dachshund observations destroy the generality of the suggestion. The only possibility worthy of consideration is that the umbrous polygenes *per se* might interact phenotypically with a^t to produce a dark sable when heterozygous $A^y a^t$, but a light or medium sable when homozygous $A^y A^y$, but even this is doubtful.

The saddle gene a^{sa} produces a characteristic V-shaped area of dark pigmentation on each side of the body, as found, for example, in the Airedale Terrier or Beagle. Superficially, the phenotypes of a^{sa} and a^t resemble each other, but saddle individuals usually display more yellow than the black and tan, especially on the face, shoulders, hind-flanks and legs. The amount of variation is such, however, that dark saddle may come to resemble black and tan. Of the two phenotypes, saddle displays the greater variation.

In the typical black and tan, as found in the Dobermann and many other breeds, the black area covers all of the dorsal surface of the body, the yellow (tan) being confined to the inside of the legs, chest markings and undersurface of the muzzle. Two characteristic tan spots occur just above the eyes.

An interesting difference between the saddle and black and tan is that the former changes in colour from birth to maturity. Saddle pups may be so dark as to resemble black and tan but lighten perceptibly with age until the typical pattern emerges. It is intriguing that sables undergo a similar transformation of lightening with age. Even A^y yellow dogs may show appreciable dark hairs in the baby and juvenile coats. The process is quite general in that even wolf cubs possess darker coats than adult animals.

The absence of the agouti A allele from most breeds, with the exception of Eskimo and Scandinavian breeds, is intriguing. The absence is so marked as to appear deliberate. Two explanations are usually offered. One is that breeders objected to the colour and wanted to distinguish their dogs from prowling wolves. The other is that mutant coat colours can be used to represent domesticity. They may be used as a hallmark of a domestic animal. This is a more traditional belief among livestock breeders than may be thought. However, there are two methods of losing the wild wolf-grey colour. This can be achieved by (1) eliminating the A gene, as noted above, or (2) by changing the phenotype of A.

In other words, one of the supposedly mutant alleles could be the A allele. If this is so, the most likeliest candidate is the a^{sa} allele. The basic pattern of A and a^{sa} is similar and the modification would be that of a darkening of the mid-region to form the saddle and a deepening of the yellow areas. It is interesting that in mongrel dogs, a transition of colours can be seen from wolf-grey to saddle. The gradual change is most noticeable for the yellow pigmentation, but is also evident in the gradual appearance of a dark more solid saddle. Some evidence that wolf-grey and saddle patterns may be identical is given by Fox (1978). A mating between a coyote and a Beagle dog gave four F_1 and $12F_2$, which are described as dark sable in the summer and light sable in winter. Fox uses the designation of sable, but the typical Beagle is saddle pattern and the illustration of the Beagle grandparent by Fox is suggestive of a dark saddle. The illustrations of the F_1 and F_2 are suggestive either of wolf-grey or of a light saddle pattern. These observations are not conclusive by any means, but they are indicative.

Little (1957) has speculated that saddle and black and tan are due to the same allele, the difference in appearance being attributed to modifying genes. It is doubtful if this speculation is true, for Willis (1976) has shown that a^{sa} is readily distinguishable from a^t and is dominant to it. Yet another speculation, that more than one a^{sa} allele exists, to explain the wide variation of expression of saddle, is also probably untrue. The variation is continuous from the darkest to the lightest animals with no sharp jumps of expression, as must be demonstrated if the hypothesis of more than one a^{sa} allele is to be taken seriously.

That a genetically different form of solid black from that produced by the A^s allele may exist in the German Shepherd dog has been known from the early observations of Yentsen (1965). These observations are suggestive, but possibly not conclusive, that black in the breed is inherited as a recessive to black and tan. Willis (1976) has tabulated breeding data which substantiate the recessive behaviour of black to both saddle and black and tan. More recently, Carver (1984) has seemingly demonstrated that black in the German Shepherd dog is inherited as a recessive to all of the agouti alleles. It would seem that the black phenotype is due to a recessive non-agouti allele symbolized by a. That such an allele should exist is not so surprising, since recessive non-agouti genes are to be found in many species of animals. What is remarkable is that both a dominant and recessive black should be present in the same allelic series! It may be cautioned that, to present date, the a allele has not been definitely detected in any breed other than the German Shepherd dog — but see comment in the section on the Belgian Shepherd dog.

Extension Alleles

The designation of this series of alleles may seem strange, but it is based upon the concept that they, like the *A* alleles, are concerned with the distribution of black and yellow pigmentation. The alleles are conceived as controlling the extension or non-extension of black pigment throughout the coat.

Three alleles are known for the extension locus:

Designation	Symbol
Brindle	E^{br}
Normal extension	E
Non-extension (yellow)	e

The normal gene is E, which is responsible for the normal extension or production of black pigment in the coat, as found, for example, in the solid black, saddle or black and tan.

The other two genes are mutant alleles of E. E^{br} is responsible for the brindled effect of black and yellow, a well-known variety in many breeds. As the symbol implies, it is dominant to E. The e allele causes the non-extension of black throughout the coat, producing a yellow phenotype. Hence, the common designation of yellow for e. Another way of imagining e is that it changes all black pigment to yellow. It is interesting that only the pigment in the hair is affected. The black pigmentation of the nose leather, lips and mouth, and rim of the eyelids are unaffected.

Little (1957) has published conclusive evidence that brindle is dominant to normal extension E instead of being recessive to it as assumed in the past. This fact justifies the use of the symbol E^{br} in preference to the older symbol e^{br}. Continued use of the latter would be misleading.

The dominance relationship between the three alleles has not been precisely worked out. Provisionally, it may be assumed that each allele is completely dominant to that lower in the series. The only reason for thinking that dominance may not be complete is the wide variation in the expression of brindle. This may vary from a blackish or dark brindle, lightened by only a few yellowish stripes, to a sandy or light brindle, lightly patterned by dark stripes. Much of this variation could be due to modifying polygenes which are independent of E^{br}. On the other hand, the average amount of brindling shown by E^{br} E^{br}, $E^{br}E$ and $E^{br}e$ could differ. In particular, the $E^{br}e$ genotype could be that with least black marking. It may be difficult to disprove this possibility in view of the overall variation, but it should be kept in view.

All fawn, yellow, orange or red animals are frequently referred to as yellow in genetic coat colour terminology, regardless of the actual depth of colour. The reason is that the coat consists entirely of yellow pigment, as opposed to black; or black and yellow, as in the saddle and tan patterns. The actual depth of colour is due to modifying genes, independent of either A^y or e. It is worthwhile to stress this point.

Little (1957) has featured the mask character as a dominant allele of the E series. He describes mask as a "superextension of dark pigment", with the symbol E^m. However, there is no evidence in the literature to support the contention. At this time it would be wise to regard the suggestion with suspicion. The heredity of mask will be discussed in a later section.

Interaction of A *and* E *Alleles*

The importance of these two series of alleles will be brought out in this section. Between them, the two series control the appearance of the black and yellow pigment in the coat of the overwhelming majority of breeds. It was necessary initially to discuss each one separately in order to describe the mutant alleles belonging to each series. Now they will be discussed together, because of the alleles interaction to produce many well-known colours.

The expression of the agouti alleles is dependent upon the presence of E, the gene governing the normal extension of black pigment throughout the coat. Their geneotypes, therefore, may be written as:

Colour	Genotype
Solid black	A^s–E–
A^y yellow	A^y–E–
Saddle pattern	a^{sa}–E–
Tan pattern	$a^t a^t E$–

When E has mutated to e, all of the pigment in the hair is changed to yellow or red. The agouti alleles cannot find expression, because their expression is dependent upon the presence of black pigment. Hence, all of the above genotypes are yellow with the substitution of ee for E:

Colour	Genotype
e yellow	A^s–ee, A^y–ee a^{sa}–ee, $a^t a^t ee$

The dog has the distinction of having two genetically different sorts of yellow phenotypes. These will be described as A^y and e yellow, as the case may be. In general, each of the genetic types occur in different breeds and each behaves as a recessive to black. That is, the A^y yellow assorts from matings of A^sA^yEE animals to each other, and *e* yellow from matings of *Ee* to each other, in the usual expectation of 3:1. The A^y yellow can be distinguished from *e* yellow when a saddle or black and tan (a^ta^tEE) is produced by two yellow parents, as occurs frequently in Dachshunds.

Many A^y yellow display variable amounts of black pigment in the coat, usually as darkly tipped hairs on the head, along the spine, on the shoulders and flanks. When the hairs are plentiful, the sable is produced. The mask pattern of black on the muzzle and ears can be shown only by A^y yellows. In contrast, *e* yellows display little signs of black tipping to the hairs or a mask. However, some A^y yellows can be so devoid of black hairs as to be indistinguishable from *e* yellows.

The discovery that both sorts of yellow may occur in the same breed (or crosses between breeds) follows from the breeding of solid black from a mating of two yellows. For example, black pups would be expected from mating a yellow of genotype A^yA^yEE to a yellow of genotype A^sA^see. The offspring will be A^sA^yEe and be black because A^s and E are dominant to A^y and *e*, respectively. A^y yellows cannot possess A^s, hence the allele must have been transmitted by the *e* yellow, thus indicating that the two yellows are genetically different.

It may be of interest if it could be determined unequivocally whether yellow dogs of a breed are either A^y or *e*. This is a difficult question to answer, since no systematic study has been made of the problem. Such information as available has come about by observations of chance matings. However, several guidelines may be promulgated. It would seem that the majority of yellows are A^y. This may be assumed for any breed in which the yellow variety displays a mask or has sable suffusion. In fact, any yellow which has a fair amount of black hairs on the ears or along the spine is more likely to be A^y than *e*.

The most exact indication for *e* yellow is where saddle pattern or black and tan bred together have produced yellow pups. Their genotypes must be a^{sa}–Ee or a^ta^tEe, respectively. Another good indication is where black pups are produced from a yellow mated to either saddle or black and tan. The yellow parent must be A^s–ee to give this result. Another indication is where black pups are produced from a mating of two yellows, as described above. The problem is then to decide which yellow is which, i.e. which is A^y and which is *e*.

Little (1957) has discussed this question of sorting out which yellow varieties of breeds could be either A^y or *e*. He has suggested that the following breeds may be regarded as being of genotype *ee*:

Beagle	Golden Retriever	Labrador Retriever
Dalmatian	Gordon Setter	Pointer
English Setter	Irish Setter	Poodle

Burns and Fraser (1966) opine that the golden Cocker Spaniel may be *ee*, but Little (1957) has suggested that both A^y and *e* yellows may be present in the breed. The presence of both A^y and *e* genes in the same breed could cause confusion, especially to a person with some genetic knowledge, but who has not realized that two different yellow genes are involved. Little (1957) has proposed that the following breeds may also possess both A^y and *e* yellows: Chow Chow, English Spaniel and Field Spaniel. The conclusions of this paragraph should be accepted with varying degrees of diffidence. Reading the literature, one can accept that some yellow varieties are *ee* with fair confidence, but others only tentatively.

The brindle allele E^{br} can induce its characteristic brindling of black and yellow only in the yellow areas produced by the agouti alleles. The action of the allele, therefore, could be visualized as inducing black pigmentation, in the form of bars or brindling, upon a yellow background.

Colour	Genotype
Solid black	$A^s\text{–}E^{br}\text{–}$
Brindle	$A^y\text{–}E^{br}\text{–}$
Dark brindle	$a^{sa}\text{–}E^{br}\text{–}$
Black and brindle	$a^t a^t E^{br}\text{–}$

The A^s gene is epistatic to E^{br}, a fact which should not be surprising since A^s individuals are solid black, with no yellow areas upon which E^{br} can find expression. In combination with A^y, however, a brindle pattern is produced throughout the coat. The $A^y\text{–}E^{br}\text{–}$ genotype is unquestionably the brindle found in most breeds. A sable A^y yellow is probably a dark brindle, due to the sable suffusion darkening the yellow areas of the brindle pattern.

The saddle brindle ($a^{sa}\text{–}E^{br}\text{–}$) displays brindling all over the yellow areas of the saddle pattern. The nature of the brindle will depend upon the type of saddle pattern. If it is a light saddle, the brindling could appear on most parts of the body, although the brindle, *qua* brindle, will be a dark brindle. If it is a dark saddle, the amount of brindling will be restricted to the areas of the region of the legs and stomach, and the dog will appear as a very dark brindle.

The black and brindle is a black and tan with the brindling confined to the normally yellow areas of the tan pattern. The phenotype of these animals is the clearest evidence that the E^{br} gene can only find expression in the yellow areas produced by the agouti alleles. These dogs are unmistakably black and tan in spite of the brindling. The amount of brindling may vary from light to dark.

Brown

The normal pigments found in the coat of the dog is black and yellow. The pigment is present in the hairs as exceedingly tiny granules. It is the colour and shape of these which give the hair its colour. In the black areas of the coat, the granules are oval and intensely brown, while in the yellow or reddish areas, the granules are smaller, round and yellowish. In liver- or chocolate-coloured individuals, e.g. chocolate and tan, the granules in the chocolate areas are a lighter brown than those found in black hairs, while the yellow granules are unchanged. Evidently, the mutant gene producing chocolate acts only upon the black pigment granules, lightening their colour. The effect upon the human eye is to change black hairs to chocolate.

The difference in colour is due to two genes, *B* for black pigment and *b* for brown pigment. The normal gene is *B* and *b* is a mutant of it. Brown is the designation used by geneticists for this type of mutant, although dog breeders will probably feel they are liver or chocolate, since the colour is a rich dark brown. The effect of *b* is to change all of the normally black areas of the animal to chocolate, not only the hair but also the skin of the lips, mouth and paws. The iris of the eye becomes a lighter colour.

The ordinary chocolate phenotype is produced by the combination of *b* with the solid colour producing gene A^s of the agouti series, as in the genotype A^s–*bb*. A fuller discussion of the various chocolate varieties will be presented in the following section in the dilution gene.

If the *b* gene has no effect on yellow pigment, it follows that there can be two kinds of yellows. "Black" yellows carrying the *B* gene of genotypes A^y–*B*– or *B*–*ee* and "brown" yellows carrying the *b* gene of genotype A^y–*bb* and *bbee*. These yellows are not completely identical, for the effects of *B and b* can be observed on the pigmentation of the nose leather, lips and inside of the mouth. These parts are black in *B* yellows but liver coloured in *b* yellows. A difference in iris colour of the eyes is also apparent.

The idea that chocolate matings could produce black offspring would be held to be unlikely in canine genetics. However, Frankling (1971) has reported five all-black litters from matings between seven different chocolate Dalmatians. She comments that the owners are most reliable and the matings were supervised. If misalliance is excluded, the other possibility is that two mutations at independent loci can produce a liver phenotype. Such a situation is extremely rare in mammalian genetics and is only known for the American mink. For this reason, Frankling's data has not received general acceptance. The possibility of two liver mutants in the dog should certainly be noted, but only on a tentative basis.

Dilution

The pigment granules in the normal hair are disposed regularly as it grows, except at the base, where the disposition tails off and the colour becomes less intense as a consequence. For example, black hairs are intensely black at the tip, but pale slightly and ultimately become bluish at the roots. The effect is due to less granules in the hair near the skin. The same effect is evident for yellow hairs. These are intensely pigmented at the tip but pale to cream towards the roots. The top colour of the coat is the intensely coloured tips and is the colour normally presented to the eye. The paler region below the tips is the undercolour. This is not normally visible, but can be exposed in long-haired individuals or if the coat is partially clipped.

The "blueness" of blue coloured varieties is also due to a lack of pigment granules, but the sparsity is caused by a different mechanism. The granules are not deposited into the hair in a regular manner but in fits and starts. Moreover, the granules may occur in clumps. The outcome is that sections of the hair may have more than the normal quota while other sections may have less or none at all. To the human eye, a genetically black coat composed of such hairs will appear slate blue, and a yellow or red coat a dull cream. The abnormal disposition of granules may be seen at high magnification under a microscope. The difference of pigment granule disposition is due to the *D* genes for normal disposition and to *d* for abnormal disposition. The genetic designation of the two are *D* for dense and *d* for dilute pigmentation, alluding to the effect presented by the coat to ordinary inspection.

There are four fundamental colours in mammals, namely black, blue, chocolate and lilac. The dog possesses all four colours and these are produced by combinations of *b* and *d* genes in conjunction with A^s and *E* as under:

Colour	Genotype
Black	A^s–B–D–E–
Blue	A^s–B–ddE–
Chocolate	A^s–bbD–E–
Lilac	A^s–$bbddE$–

The above are the four basic colour varieties to be found in many breeds. The same colours may be found combined with a^t to give the various bicolours or tans:

Colour	Genotype
Black and tan	$a^t a^t B\text{–}D\text{–}E\text{–}$
Blue and tan	$a^t a^t B\text{–}ddE\text{–}$
Chocolate and tan	$a^t a^t bbD\text{–}E\text{–}$
Lilac and tan	$a^t a^t bbddE\text{–}$

The *d* gene affects both black/brown and yellow pigmentation, hence the blue and tan and lilac and tan should be more correctly described as blue and cream and lilac and cream, respectively. However, the "and tan" designation may be used in a formal manner of speaking to denote the tan pattern, so long as it is recognized that the tan is diluted to a cream for the blue and lilac.

The solid black and black and tan are the same colours as discussed earlier, except for the addition of genes *B* and *D* in the genotype. This is necessary to indicate how the genotypes differ from those for blue and chocolate. Gene *d* is present in a great many breeds, the proportion depending upon the popularity of the blue variety. Gene *b* is not so common, but it occurs particularly among the Spaniels. Some Spaniels, indeed, can only be homozygous *bb* in constitution, such as the American and Irish water and Sussex, while others permit either *B* or *b* dogs. It is curious that breeds which have *b* do not as a rule have *d*, hence the lilac combination *bbdd* never or rarely makes an appearance. However, this genotype forms the basis of the unique colouring of the Weimaraner breed, with the genotype $A^s\text{–}bbddE\text{–}$. Most fanciers have their preferred terms for the various colours. Canine breeders are no exception, in that liver is frequently used to describe the chocolate. The lilac colour is so uncommon that no general canine term exists. The colour of the Weimaraner is described as silver-grey, which admittedly is an apt description. However, lilac is a term in common usage by mammalian colour geneticists for *bbdd* animals, and dog breeders would be wise to be aware of it.

The two genes will produce four sorts of brindle-coloured dogs by a similar combination of *b* and *d* genes in conjunction with A^y and E^{br}:

Colour	Genotype
Black brindle	$A^y\text{–}B\text{–}D\text{–}E^{br}\text{–}$
Blue brindle	$A^y\text{–}B\text{–}ddE^{br}\text{–}$
Chocolate brindle	$A^y\text{–}bbD\text{–}E^{br}\text{–}$
Lilac brindle	$A^y\text{–}bbddE^{br}\text{–}$

The above remarks apply to all genetic forms of brindle discussed previously. It is doubtful if some of the chocolate and lilac forms would be readily recognized for what they are. In any event, they would be uncommon.

The b gene acts only upon black pigment, hence all yellow/red phenotypes look alike as far as hair pigment is concerned, but the presence of the gene is revealed by a change in skin and iris colour. The d gene dilutes the yellow/red coat colour to a cream shade, with but a small change of either skin or iris colour.

The sable has the genotype $A^yB–D–E$ plus a variable amount of black suffusion. The latter would be modified to chocolate by the action of b, to create a chocolate sable. However, it is doubtful if these would always be readily recognized as such. Many breeders would probably classify them as yellow or red, especially if the sable shading was light to moderate, because of the lack of contrast between the chocolate tipping and the yellow/red background. Similarly, the blue or lilac sable would not be immediately recognizable from ordinary cream. The effect of the sable polygenes would be to give the creams a bluish suffusion.

Hair colour	Skin colour	Iris colour	Genotypes
Yellow/red	Black	Brown	$A^y–B–D–E$ $A^s–B–D–ee$ $A^y–B–D–ee$ $a^ta^tB–D–ee$
Yellow/red	Liver	Light hazel	$A^ybbD–E–$ $A^s–bbD–ee$ $A^y–bbD–ee$ $a^ta^tbbD–ee$
Cream	Slate	Brown	$A^y–B–ddE–$ $A^s–B–ddee$ $A^y–B–ddee$ $a^ta^t–B–ddee$
Cream	Pale liver	Light hazel	$A^y–bbddE–$ $A^s–bbddee$ $A^y–bbddee$ $a^ta^tbbddee$

Albino Alleles

The albino series are the fundamental controllers of pigment production throughout the coat. It may be wondered why the series are called albino, since albinism is extremely rare in dogs. The answer resides in general mammalian colour genetics. Albinism may be rare in dogs, but it is common in many other mammals. The albino locus, in fact, is more mutable than most, and has produced a series of alleles with characteristic phenotypes. Further, these phenotypes are similar in all species, so that information for one holds good for other species with small risk of error. This is the situation for the dog, where the amount of conclusive breeding data is so scanty.

The total number of mutant alleles of the albino series is unknown, but at least three may be regarded as established:

Designation	Symbol
Full colour	C
Chinchilla	c^{ch}
Blue-eyed albino	c^{b}
Albino	c

All dogs with normal expression of coat colour have the *C* gene. The gene is designated as full colour for the simple reason that it permits the full expression of colour. It is typical of albino alleles that each one down the scale allows the expression of less and less pigment, yellow pigment being the first to be degraded, followed by black. The final step is total absence of pigment, as exemplified by the albino.

Chinchilla alleles, c^{ch}, characteristically change yellow to a pale fawn colour, while either not changing black or changing it very slightly. Such genes are likely to be found in breeds with pale yellow pigmentation, such as the Norwegian Elkhound. The effect of c^{ch} is most noticeable for yellows of either A^{y} or *e* genotype. In particular, the pale yellow or creams of the Golden Retriever would correspond almost exactly to the phenotype expected for $c^{ch}c^{ch}ee$. The colour of these chinchillated yellows (to use their technical name) can vary appreciably, from a warm cream to almost white. Much of this variation could be due to rufus polygenes. The wide variation has prompted Little (1957) to suggest that more than one chinchilla allele may be present in dogs.

This second allele (symbolized as c^{e} by Little) could be responsible for the almost white phenotype with pale cream suffusion along the spine, shoulders and head. If such an allele exists, the genotype would be either A^{y}–$c^{e}c^{e}$ or $c^{e}c^{e}ee$. White varieties of a few breeds could conceivabely have this genotype; for instance, as Little (1957) has suggested for the West Highland White Terrier. None of the extremely pale phenotypes are albino, because they have dark eyes. It may be mentioned that the genetic basis for even one c^{ch} type allele rests mainly upon phenotypic observations, rather than upon breeding experiments.

The speculation that c^{ch} alleles exist would be legitimate but for the probability that rufus polygenes with minus properties can dilute yellow pigmentation to a cream, in essence mimicking the expected effects of c^{ch}. For this reason, only one chinchilla allele should be recognized at this time and that only provisionally. It may be noted that Little (1957) frequently mentions in his book that *C* is incompletely dominant to c^{ch}. This suggestion is at variance with the behaviour of c^{ch} mutants in other species, where complete recessiveness is the rule. If minus rufus polygenes are at work diluting yellow rather than c^{ch}, the apparent incomplete dominance of *C* would not be surprising.

Pearson and Usher (1929) have described two very light coloured phenotypes, each possessing eyes with pale blue irises and dull red pupils. More than one writer in canine genetics has assumed that one of these is an albino allele, occupying a position relatively low on the scale of pigment production, in fact probably not far removed from complete albinism. The allele has been designated blue-eyed albino (c^b). Unfortunately, little is known about the heredity of c^b, but it is reasonable to place it just above c in the series.

Complete albinism, as represented by white coat, translucent irises, pink pupils and skin, may be very rare, but, nevertheless, animals of this type have been reported by Whitney (1947) in the Pekinese and other breeds. Little (1957) has also reported albinism in Pekinese. The variety is probably rare because of a general dislike of the colour by breeders, although there is no reason why albinos should not be bred in non-working breeds.

Pink-eyed Dilution

This form of dilution is common in mammals in general, but is rare in dogs. However, it should be considered because it is probable that one of the pale coated and bluish-red eyed phenotypes described by Pearson and Usher (1929) was of this genetic type. It is typical for this type of dilution to produce a bluish or greyish coat colour and reddish eyes. In this latter respect, pink-eyed dilution differs from ordinary dilution, where the eyes remain dark. The red eyes are not usually completely devoid of pigment (otherwise they would be pink) and the iris is often bluish (not translucent as in the complete albino). Despite the reduction in eye pigmentation, pink-eyed dilution has no connection with albinism. The symbol for pink-eyed dilution is p, the gene being a mutant of P, the normal gene producing an intensely coloured coat and dark eyes.

The reason for thinking that Pearson and Usher have discovered a pink-eyed type of mutant is that matings between some of their "albino" dogs produced pups with dark coloured coats and eyes. Apart from accidents of mating, this result could occur only from crosses involving two independent recessive genes, each producing red eyes and pale coats. Thus, in present terminology, the crosses could have been of the category pink-eyed dilution ($CCpp$) by albinotic (c^bc^bPP) giving black offspring (Cc^bPp). It may be noted that Pearson and Usher compare their "Cornaz albino" with pink-eyed lilac mice. These mice are known to be non-agouti pink-eyed dilute of genotype $aapp$; the corresponding genotype for the dog would be A^sA^spp, gene a producing a solid black in mice, exactly as does A^s in dogs. It is rather unfortunate that Pearson and Usher wrote as if all of their red-eyed dogs were albinos, which they were not.

At this time it is impossible to be absolutely sure of the identity of the pale coloured dogs described by Pearson and Usher. If these investigators had discovered a pink-eyed dilution mutant, an interesting range of coat colours would have been possible. The mutants A^s and A^y , and the probable mutant c^b , were present in the stock, from which the following combinations are possible:

Coat colour	Eye colour	Genotype
Black	Black	A^sCP
Yellow	Black	A^yCP
Black/brownish black	Red	A^sc^bP
Cream/off-white	Red	A^yc^bP
Slate grey/lilac grey	Red	A^sCp
Yellow	Red	A^yCp
Off-white	Pink	A^sc^bp
White	Pink	A^yc^bp

The genes c^b and p would interact with each other in the genotype $c^b c^b pp$ to produce a pseudo-albino or mock albino, each acting to remove such pigment as left by the other. The eye would be pink and the coat completely white or only slightly off-white. If *b*, the gene for brown pigmentation, is added, the effect on eye colour could have been even more marked, for the combination bbc^bc^b *or* $bbpp$ would probably have light eyes, regardless of any further effect *b* may have on coat colour. That is, c^b *and p* may have had a greater effect on eye colour of *bb* animals than upon *B* animals. Pearson and Usher did not consider any of the above interactions because they were not in the mainstream of mammalian colour genetics and they regarded all red- and pink-eyed phenotypes as albinos. If one wishes to be precise, this assumption is simply not true.

Slate-grey

Little information is available on the slate-grey gene, except that it is inherited as a dominant and is phenotypically similar to blue dilution. The gene (symbol *Sg*) has been reported only in the Rough Collie by Ford (1969).

Powder Puff Dilution

This is an unusual dilution in that it is largely transitional. The baby coat of mutant black pups is grey in colour, but changes gradually to black. By 6–8 months, the coat is normal except possibly for a slight dilution and a pale undercolour. In black and tans, the yellow areas are not diluted. The nose leather is unaffected. The powder puff gene (symbol *pp*) has been reported only in the Rough Collie (Lund *et al.*, 1970). The incident of the gene is unknown.

Merle

Merle is a name given to a sometimes mingled, but more often patchwork, combination of light and dark areas, such as occurs in the Rough and Smooth Collie, Shetland Sheepdog, Dappled Daschshund and Harlequin Great Dane. The Merle effect is produced by heterozygosity of a gene M which is dominant to normal colouring. The light areas are produced by a mixture of normal and pigment-deficient hairs, the black/chocolate pigments being more affected than the yellow. The homozygote MM has a solid white coat, partially or completely blue irises, smaller eye bulbus, and is usually partially or wholly deaf and often sterile.

The genotypes of the common Merle colours are as follows:

Name	Genotype
"Blue" merle	$A^s–Mm$
Blue/bicoloured merle	$a^t a^t Mm$
Yellow/sable merle	$A^y–Mm$

The "blue" or dapple merle offers the greatest contrast between the normal and bluish merle areas. Indeed, the typical animal appears as a bluish dog with irregular or ragged black spots and patches of varying sizes. A similar effect may be seen in the black and tan merle, at least upon the back but less so on the stomach. This dual effect raises an interesting point. The M gene does not appear to have such a diluting effect on yellow as upon black pigmentation. Thus, in ordinary yellows the patchwork merle pattern is not so evident. This lack of contrast does not apply to sables, depending upon the amount of sable suffusion, because the sable colour is changed to bluish. The M gene increases the amount of white in white marked animals. Such animals are often termed tricolours. It is patently unwise to breed the white homozygous merle MM, hence it is recommended that all matings should be between merle heterozygotes (Mm) and normal (mm) animals. The expectation is half merle and half normal puppies.

Harlequin

The Great Dane has a fascinating colour variety known as the harlequin. The coat is white or off-white, liberally sprinkled with ragged black patches of varying sizes. That the colour has a resemblance and a close genetic connection with the merle coloration has been appreciated for many years but only comparatively recently has the connection been fully understood (Sponenberg, 1985; O'Sullivan and Robinson, 1989). While the merle is a patchwork of black and slate-bluish areas, the harlequin is a patchwork of black and white, the admixture being more apparent for the latter because of the greater contrast between the colours. The harlequin is basically a merle but with a modifying gene *H* which changes the bluish areas of the merle to white.

The genotypes of the merle and harlequin may be shown as follows:

Name	Genotypes
White merle	*MMHh, MMhh*
Merle	*Mmhh*
Harlequin	*MmHh*
Normal	*mmHh, mmhh*

The *H* gene may be regarded as a modifying gene of the merle coloration since it only changes the appearance of this phenotype. The gene cannot find expression with either white merle or with black. The former is perhaps not remarkable since the white merle is without pigment but the lack of expression for the latter deserves to be noted. Moreover, detailed examination of breeding records revealed that the *HH* individual dies at an early age, probably in the uterus. As a consequence, it is only possible to have the six genotypes listed above. This is not all, because the records indicate that a proportion of white merle of genotype *MMHh* die due to the effects of the *H* gene. Normally, white merles may have defects of vision and hearing but they are not regarded as having a high death rate.

Tweed

This colour known as tweed in the Australian Shepherd dog is also based on the merle gene *M*. The usual merle coloration is a patchwork of black and slate blue, the latter varying in intensity but usually consistent throughout the coat. While the tweed coloration displays the typical merle patchwork, the bluish areas are of different intensities. Each patch is of uniform intensity but a dog may have as many as three or more patches of different shades. The patches appear to be more clearly defined than those expressed by the ordinary merle. Tweed has been shown by Sponenberg and Lamoreux (1985) to be due to a modification of the typical merle phenotype. The responsible gene is designated as tweed and symbolized by *Tw*. The genotypes of the merle and tweed colours may be represented as follows:

Name	Genotypes
White merle	*MMTwTw, MMTwtw, MMtwtw*
Merle	*Mmtwtw*
Tweed	*MmTwTw, MmTwtw*
Black	*mmTwTw, mmTwtw, mmtwtw*

The *Tw* gene is inherited as a dominant to normal colouring but is only expressed in conjunction with the merle phenotype. The gene, therefore, is a modifier of merle, similar to *H*, but apparently less extreme in effect. That the *Tw* gene cannot find expression in white merle is to be expected since the phenotype is devoid of pigment. However, the *Tw* gene cannot find expression in the non-merle of black phenotype. This latter finding defines *Tw* as a true modifying gene.

CN Dilution

This gene dilutes both black and yellow pigment, the black to a dull grey and the yellow to light beige or off-white. The sable is changed to a silvery grey to almost white, according to the amount of sable shading. The nose leather is coloured light tan, a conspicuous feature which differentiates CN dilution from the other dilute phenotypes. The hair of CN pups is of a finer texture than normal, and may appear slightly wavy, but this effect disappears in the adult. The gene is inherited as a recessive (symbol *cn*) and is semi-lethal. The gene causes a periodic deficiency of neutrophiles in the blood, which seriously affects the ability of the individual to resist bacterial infection. The majority of CN pups die within a few months of age (Ford, 1969; Lund *et al.*, 1970). CN dilution has been only definitely identified in the Rough Collie.

Progressive Silvering

Little (1957) has proposed that the progressive silvering or greying which is present in some breeds is due to an incompletely dominant gene *G*. He gives no breeding data to support the contention, but Whitney (1952, 1958) has published data which indicates that progressive silvering is monogenically determined. Whitney initially proposed *si* for the gene, but later adopted Little's *G*. The expression of the character is variable, and a most experienced eye is often necessary to identify all heterozygotes. It is a moot point whether or not the gene should be regarded as a semi-dominant or a recessive. This aspect could well be a breed or strain characteristic.

Silvering in mammalian genetics is a term used to denote a coat with abundant white hairs. If the number of black hairs outnumber the white, so that the animal appears as a black ticked with white hairs, the condition is known as silvered. If the number of white hairs so outnumber the black that the animal appears white ticked with black hairs, the condition is known as roan. These terms could describe the expression of two different genetic forms of silvering or different grades of expression of the same character. Silvering is also classified as stationary or progressive; stationary when the silvering occurs at a definite stage of coat development, then remaining unchanged to any significant extent, and progressive when the silvering increases steadily over a definite period or even throughout life. The silvering associated with *G*, therefore, is of the progressive type.

Puppies carrying *G* are born black but develop a progressively greying of the coat. The heterozygote *Gg* may change to a slaty-blue, fairly quickly in a few cases but slowly over several years in the majority. A similar change occurs for homozygotes *GG*, but more drastically. Commencing at a few weeks of age, the coat is transformed into a silvery-blue by the time of maturity. The expression is variable, depending upon breed as much as the individual dog. The colour may be uniform or vary over the body. In Bedlington Terriers, the top of the head and parts of the shoulders may be pale to almost white. In Poodles, Whitney (1958) considers that the off-black or taupe colour is an expression of the heterozygote *Gg*, maintaining that these change gradually to a dark bluish-grey over some 3–5 years.

In combination with other mutants, *Gg* produces a small dilution of coat colour while *GG* produces a sharp reduction. The light phenotype of *GG* implies that it is almost but not quite epistatic to other colour genes. The following genotypes are probably the most interesting:

Colour	Genotype
Silver	A^s–*B*–*D*–*E*–*GG*
Chocolate silver	A^s–*bbD*–*E*–*GG*
Bluish silver	A^s–*B*–*ddE*–*GG*
Cream silver	A^s–*B*–*D*–*eeGG*
Light cream silver	A^s–*B*–*ddeeGG*

All of the above are silver or roan animals, with abundant white or light hairs intermingled with the coloured. If the silver has many coloured hairs, the underlying phenotype is usually apparent, but if the silver is almost white, identification may have to rely upon a general impression. All *bb* can be identified by liver nose leather and light eye colour; the coat may show a light brownish tinge. Similarly, the *ee* silvers will probably show a pale yellow or cream tinge. All of the phenotypes would be easily identified before the silvering develops. Note that A^s–*B*–*ddE*–*GG* would have a blue coat even before the silvering developed, and this is due to the dilute gene *d*, not *G*. The expression of silver may be pale in this genotype because of the effect of *d*.

Any breed possessing a variety which is born with a normal colour but then pales steadily to an appreciable lighter colour could be *GG*. It is reasonably certain the *G* gene is present in the Bedlington Terrier, Kerry Blue Terrier, Old English Sheepdog and Poodle. Little (1957) has suggested that the gene may be present in the Cairn Terrier, Dandie Dinmont, Skye Terrier and Yorkshire Terrier, but direct evidence is lacking for all of these breeds. Several other breeds could be usefully investigated for the presence of *G*, among them the Australian Terrier and the Australian Silky Terrier.

White Pattern

The presence of a white pattern is characteristic of many dog breeds. In many other breeds, furthermore, the presence of white markings is accepted, even if it is not the dominant feature. This implies that the white pattern must be considered as an integral part of the genetic make-up of the dog. Genetically, the white areas are referred to as white spotting, regardless of the amount of white. The term actually grew out of situations where the amount of white was limited and did occur as spots (e.g. in mice). Now the term is used more generally to include situations where the amount of white is extensive, even including certain all-white animals. These latter may be rightly regarded as one big spot!

The distribution of white markings follows a relatively orderly progression as the amount of white increases. The first appearance of white is on the chest, feet, muzzle and end of the tail. As the amount increases, the whole chest, legs and stomach becomes white. A break of white extends over the shoulders and withers. Another break often extends up the flanks and over the back. Subsequently, there is a tendency for the pigmented areas to appear as slabs, splotches and spots on the body, as the amount of white increases further. Even at this stage there is a certain regularity of occurrence of the spots if a series of dogs is examined. The splotches and spots occur on the sides of the body, not distributed entirely at random. The last places to be depigmented are the top of the head, particularly the ear region, and at the base of the tail. Figure 29 presents a generalized picture of the progressive increase of the amount of white.

The reason for the generally ordered progression of white is due to an absence of pigment-producing cells in the skin at the root of the hairs. These cells arise at definite locations corresponding roughly to the top of the head and along the spine in the developing embryo. If there is interference, so that the number of primordial sites is reduced or the rate of spread of the pigment-producing cells over the body is retarded during development, the puppy will be born with white markings. Although the skin is normal and hairs are formed, they will be colourless (appearing as white to the human eye) because pigment-producing cells have failed to arrive at their appointed positions. Since embryonic development is an orderly process, the failure of the pigment-producing cells to take up their positions will also be orderly. Ergo, the process shows up as a more or less regular progression of a white pattern, ranging from a trace to a completely white phenotype.

It must be appreciated that despite the discernible trend in the increasing amount of white spotting, there will be considerable variation for individual animals. This is especially so for dogs with extensive white, for the pigmented patches or spots may occur anywhere on the head, shoulders, sides or flanks and certainly not necessarily symmetrically. Against this is

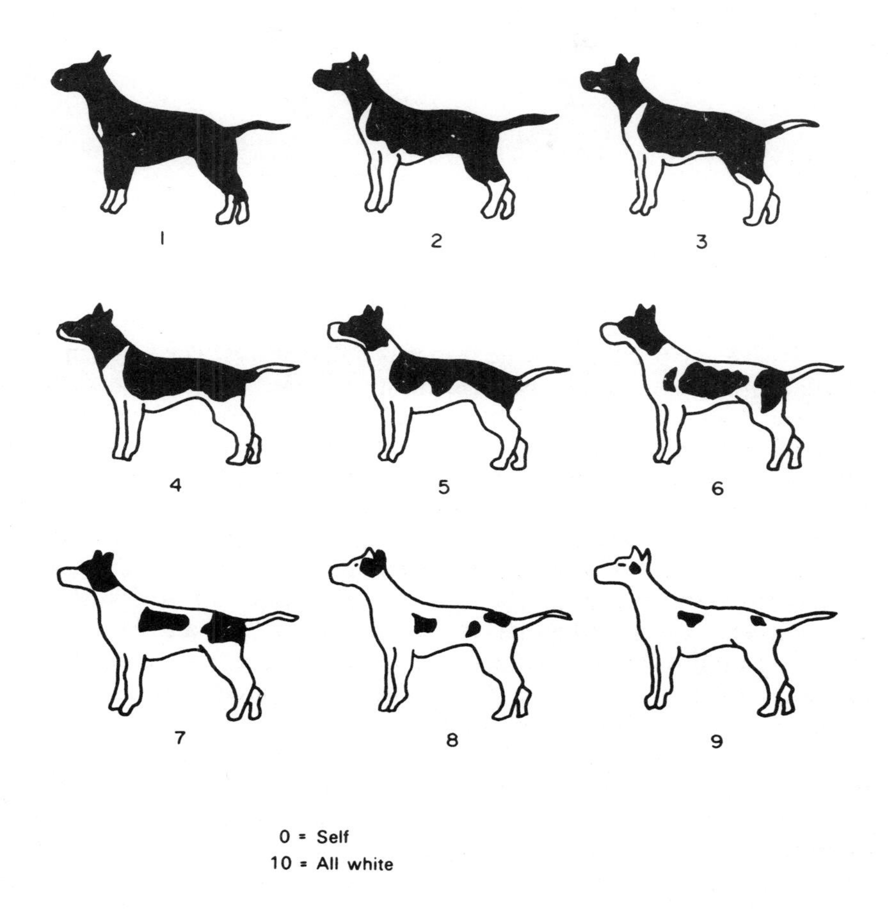

FIG. 29. The expression of white pattern shown as a series of ten grades. Grade 0 is the self or non-spotted dog while grade 10 is the all-white dog (not shown).

the relatively stable appearance of a facial stripe and narrowing streak of white extending up the front legs, shoulders and meeting at the withers, the so-called "collar" grade of white, which is a feature of breeds such as the Boston Terrier, Collie and Saint Bernard. The more spotted animals, that is pigment patches or spots on a white background, are commonly known as piebalds. Some of this variation is due to polygenes modifying the expression (i.e. amount of white) of the main spotting alleles and some due to accidents of embryonic development. This is shown by significantly less variation of expression for those breeds in which a certain white pattern is held to be desirable, but without totally eliminating the variation.

Little (1957) has hypothesized that the distribution of white spotting can be explained on the basis of three mutant alleles of a gene S for self (non-white):

Designation	Symbol	Grades
Self	S	0
Irish spotting	s^i	1–3
Piebald spotting	s^p	3–9
Extreme-white spotting	s^w	9–10

The Irish allele produces the least spotting of the three. The white markings vary from spots on the muzzle, chest, feet, etc., to that of a narrow collar encircling the neck or withers. The Basenji is portrayed by Little as exhibiting typical expression and variation of $s^i s^i$ animals.

Animals homozygous for piebald spotting possess greater amounts of white, on the average, than those homozygous for Irish spotting. However, a feature of the s^p allele is its wide variation of expression. For example, according to Little, the Beagle has the genotype $s^p s^p$, yet the majority of Beagles display a white pattern only a little more extensive to that shown by the Basenji. On the other hand, many Beagles possess a grade of spotting never attained by the Basenji, thus revealing the true nature of the spotting. At the other extreme, the Fox Terrier is stated by Little to have the genotype $s^p s^p$ and in this breed the amount of white is so extensive that it is commonly regarded as a white animal with coloured spots.

The extreme-white allele produces the greatest amount of white, resulting in an individual which is white except for a small spot or spots on the rump or head. If no spots are present, the dog is a black-eyed white. A number of breeds have such a phenotype either as the standard colour or as a recognized variety. The clue to whether a black-eyed white is due to the genotyope $s^w s^w$, or some other, is that while most individuals will be devoid of coloured spots, a small but persistant minority will exhibit spots. On this criterion, Little has shown that the black-eyed white of the Bull Terrier, Great Pyrenees, Samoyed and Sealyham Terrier is $s^w s^w$.

It is typical of spotting alleles to be incompletely dominant to each other, but for this to be erratic as far as individual animals are concerned. That is, when the average amount of white in the coat is taken, incomplete dominance is revealed, but that the variation of expression of the various genotypes will overlap each other. On this basis S is dominant to s^i but not to s^p or s^w. The heterozygotes Ss^p and Ss^w display the amount of white typical of Irish spotting. The Irish allele s^i is dominant to s^p but not over s^w, the heterozygote $s^i s^w$ approximating the piebald grade of spotting. The relationship between s^p and s^w is unclear because of lack of data, but the heterozygote $s^p s^w$ is probably a lightly spotted piebald. It should be remembered that the expression of s^p shows the greatest variation of the alleles and seriously complicates an accurate determination of the dominance relationship.

The expression or grade of white shown by all of the alleles is partially controlled by modifying polygenes. These appear to act by determining grade of white, rather than by whether the spotting should occur on any part of the body. That is, the regularity and positioning of the white is scarcely modified, regardless of the breed. The fact that breeds differ in the amount of white which they typically display is due to the particular allele or alleles which they possess and to the modifiers particular to the breed. The importance of modifiers in the determination of white markings needs stressing, since of all the colour characters white spotting is subject to the greatest modification in this respect.

Whether or not these polygenic modifiers produce white spotting in their own right or merely decrease or increase the expression of the *s* alleles is open to debate. It is almost certain that some are capable of inducing minor white spotting or low grade Irish spotting in homozygous self *SS* individuals. Little gives a list of seven breeds where this type of spotting has been observed in a small proportion of dogs. In some cases, small spots of white may arise from accidents of embryonic development, but it is doubtful if all cases can be explained in this manner. The situation is complicated by the fact that many polygenes are probably involved and that expression is also variable. That is, many *SS* animals which ought to display some white, because of the polygenes they possess, do not. In any event, persistent selection against them is the only means of keeping the frequency of minor white spotting at a low level of those breeds where it is definitely undesirable.

Both Whitney (1947) and Little (1957) have noted that the amount of white in the coat differs between black and red Cocker Spaniels. The difference is quite striking and transcends a simple comparison of average grades, for in Little's data the majority of blacks show little or moderate amounts of white, while the majority of reds show appreciable amounts. The reason for this is unknown, but could arise from a particularly influential modifier of s^p being linked to gene *e* or that pigment-producing cells containing *ee* are less able to spread throughout the skin in comparison with those containing *E*. This inability may not be apparent in *S* animals but becomes obvious when s^p compounds the inability. Burns and Fraser (1966) note that chocolate Cocker Spaniels usually have less white than red but more than black. This implicates the *b* gene as a factor in the development of the amount of white.

It may seem strange that the various white-spotting genes are alleles of a single locus. In many mammalian species, white spotting is produced by mutants at different loci. A similar situation could be expected for the dog, as mentioned by Hutt (1979). However, the data published by Little (1957), which is extensive by canine standards, is fully consistent with a single series of alleles. It is possible to quibble whether two or three distinct alleles have been isolated (sometimes it is difficult to separate s^i and s^p), but the results tend to favour the latter. It is just possible that an independent gene, with variable and minor expression, independent of the *s* alleles may exist, but the evidence is equivocal.

White

A completely white coat but with dark eyes, darkly pigmented nose leather, and light to heavy pigmentation of exposed skin areas was described by Carver (1984). The colour occurs as a variety of the German Shepherd dog and is inherited as a recessive to normal coloration. The gene is symbolized by *w*. The author considers that the gene is not s^w, but this possibility should be borne in mind.

Ticking

The white areas of many breeds, notably Pointers, Setters and Spaniels, are not clear, but covered with innumerable small pigment spots. This is known as ticking or flecking and is due to a dominant gene *T*. Ticking is not present at birth, but appears within a few weeks at the first moult. The expression is very variable, ranging from a few spots to such a profusion of spots that the white areas appear as roan. In long-haired breeds, the ticking simulates roan because of the greater intermingling of the hair fibres. Some breeds show a tendency for more frequent ticking on the legs and stomach rather than an even distribution.

Ticking is the outcome of an interaction between any of the alleles for white spotting and *T*. The usual allele involved is s^p, because this is the most common spotting and produces areas of white large enough for the ticking to be seen clearly. The ticking does not interfere with the colour genes, hence the colour of the ticking corresponds to that normally expected at any particular position. The genotypes of individuals exhibiting ticking will be $s^p s^p T-$, $s^p s^w T-$ or $s^w s^w T-$, bearing in mind that the amount of white is very variable and that ticking occurs only in the white areas. The remainder of the genotype would be that appropriate for the colour variety. For example, a ticked black has the genotype $A^s-B-D-E-s^p-T-$ and yellow ticked the genotype $A^y-B-D-E-s^p-T-$. Black and tan ticked ($a^t a^t B-D-E-s^p-T-$) have black ticking on the back and tan ticking on the stomach.

The expression of ticking varies from a lightly ticked dog with a few spots merely on the legs and stomach to the heavily spotted form found in a number of hound breeds (e.g. the Gasgons and Griffons). Here the effect is that of roan. On the other hand, the ticking on some dogs, although profuse, tends towards flecking similar to but smaller and less distinct than that shown by the Dalmatian. Hence, there is not only wide variation of quantity of ticking but also of its quality. Evidently, ticking is a malleable character. Whether or not this indicates selection and stabilization of the polygenes governing the expression of ticking has yet to be finally settled.

Flecking

The Dalmatian type of solid spotted pattern has been attributed to an independent gene in the past but Little (1957) has shown that it is a modification of ordinary flecking. The basic genotype is $s^w s^w TT$. That is, the coat is mostly or completely white due to the s^w gene and liberally covered in coloured spots due to T. However, the spots are solid in the Dalmatian pattern rather than interspersed with white hairs. It may be noted that Little has remarked that the crossing of Dalmatian with white marked dogs results in a reversion to the flecking-type of spotting. There the situation remained until Schaible (1976, 1981) discovered the presence of a gene in the Dalmatian which prevented the occurrence of white hairs in the spots. The gene is inherited as a recessive to flecking. Denoting the gene for interspersion of white hairs (flecking) by F, the gene for solid spots will be f. The full genotype for Dalmatian type spotting, therefore, is $ffs^w s^w TT$. The f gene does not seem to have any effect on phenotypes other than those containing T in combination with one of the white spotting alleles.

Mask

Mask is the term given to the suffusion of black pigment around the muzzle, often covering most of the forepart of the face and the ears. Its most obvious expression may be observed in A^y yellow dogs, in such diverse breeds as the fawn varieties of the giant Mastiff and diminutive Pug. It may be present in animals carrying a^{sa} and a^t, where it will obscure the yellow pattern associated with these genes, but is, to produce black-faced saddle and black and tan. The former may not be particularly noticeable because the presence or absence of mask could be taken as part of the variation of saddle. However, the expression of black and tan is more stable and a black and tan with little or no facial markings is immediately noticeable. Mask would not find expression in A^s animals, but these can carry the character.

Mask may vary in the extent to which it covers the face or pigments the ears, but this does not necessarily mean that it is polygenically inherited. Very little is known for the heredity of mask. Little (1957) cites the case of a litter born to a yellow Dachshund, mated to a masked fawn Pug, in which all of the eight yellow pups had black masks. These could indicate that mask is inherited as a monogenic dominant to non-mask. Provisionally, at least, mask may be attributed to a gene *Ma*.

Little (1957) depicts mask as produced by a dominant gene in the extension series of alleles. However, he does not give any date to support this assertion and there are grounds for thinking that it is false. Brindle animals may or may not possess the mask characteristic, which in itself is sufficient to cast doubt upon the assertion. Mask is probably not expressed in *ee* animals, but this is not necessarily evidence that mask is in the extension series. It is probable that *e* yellow animals cannot express any black pigment in the hair. In all, it is wise to suspend judgement that mask, if it is due to a single gene, actually belongs to an established allelic series.

The probable absence of mask in *e* yellow animals may be seen as a guide towards distinguishing between A^y and *e* yellows. All yellow animals with mask may be taken to be A^y, while those without could be either A^y, or *e*, but probably the latter.

Rufism

The intensity of yellow pigment in the coat may vary from pale cream to rich red. Some of the variation can be attributed to the action of major genes, such as dilution *d* or possibly chinchilla c^{ch}, but this is not the end of the story. There still remains appreciable variation which is due to a group of polygenes known as rufus. The effects of the group are at their most obvious in both A^y and *e* yellow animals. The intensity of yellow varies greatly in these, depending upon the individual, strain or breed. This reveals the considerable influence and importance of the group in determining the precise intensity of yellow coloration.

The influence of the group extends beyond that of modifying the colour of yellow animals, for they are also responsible for the intensity of the yellow areas of both the saddle and tan patterns, for these may vary from pale yellow to red. This raises the status of the group beyond that of simple modifiers of A^y and *e* genes, as they are sometimes portrayed, into a fundamental group determining a component of coat colour. The only varieties to escape their influence would be solid black and all-white.

The action of the rufus polygenes appears to be mainly in a plus direction, that is they deepen the colour. The wolf has a drab coloured coat and breeders of the past have evidently selected for a brighter coloured dog, judging from the appearance of the majority of breeds. Iljin (1932, 1941) found that the paler colour behaved as a dominant to the richer, and he postulates a series of these "intensification" alleles to explain the variation (symbols *Int*, int^m and *int*, respectively). The *Int* allele produces the palest, *int* the deepest and int^m an intermediate colour. It is quite possible for a major gene to be involved in the intensity of yellow, but Iljin's postulation is based upon too few observations for this to be accepted as definite. At present, it is wise to assume that the pale yellow to rich red variation is due to numerous rufus polygenes.

Iljin's observations are interesting and probably of significance, since the loss of rich yellow coloration in crosses between bright and pale coloured dogs have been observed by Tjebbes and Wriedt (1927). Burns and Fraser (1966) and Willis (1976) have commented that richness of red can be easily lost and difficult to regain once lost. These comments imply that the rufus minus polygenes may be dominant to the plus polygenes. However, it is characteristic of polygenic groups that these may show dominance in some crosses but not in others, depending upon the composition for particular parents, hence one should be on guard for exceptions to any rule which may be laid down.

Saddle Variation

The expression of saddle varies in a remarkable fashion, so much so that it has been proposed that more than one saddle allele exists. That is, for example, one allele producing a dark saddle and another a light saddle. However, this is doubtful. The variation is continuous and can be easily explained by modifying polygenes.

The variation extends from a lightly patterned saddle, which could be confused with a sable, to a dark saddle, not far removed from a black and tan. The confusion with sable, is heightened by the tendency for saddle pattern to become lighter with age. This parallelism reflects the genetic closeness of the two alleles, not necessarily that they are identical. In any event, it is only the lighter and more diffused patterned saddle individuals which are likely to be confused with sable. At the other extreme, where the saddle has spread over most of the body, the pattern comes close to being mistaken for a black and tan.

The saddle refers to the V-shaped dark pattern on the sides of the body, extending from the spine to a point roughly ending at the belly line. However, both the area and intensity of the pattern can vary. When the pattern is intense, it is fairly easy to recognize, regardless of area, but when it has been infiltrated with lighter hairs, it becomes ill-defined and less easy to recognize. The pattern does, in fact, tend to resemble the wild wolf-like pattern, and this is the basis for the suggestion that saddle is a modification of the basic agouti A allele. In this respect, the resemblance is enhanced if the yellow aspect of the saddle pattern is pale. Strictly speaking, saddle pattern only refers to the distribution of black pigmentation. The depth of colour of the yellow is governed by rufus polygenes and these are independent of saddle pattern.

It is interesting to speculate that some of the variation of saddle is due to umbrous polygenes. That is, the same group which changes A^y yellow into sable could produce the dark grades of saddle. Unfortunately, no breeding experiments have been done to investigate the possibility. The intensity of the yellow areas may vary independently of the dark saddle pattern. These polygenes, as the name implies, engender a darkening of the coat. This is almost certainly due to action of the rufus group of polygenes.

Umbrous or Sable

Their most obvious expression is to change an ordinary A^y yellow into a sable. Most, if not all, A^y yellows have some black-tipped hairs, normally on the head and along the spine. The baby coat is usually liberally covered with dark hairs, giving the pup a dark, dusky appearance. These disappear as the pup matures and an adult coat is attained.

As the number of dark hairs increases in the adult, so the sable coloration is engendered. The degree of suffusion varies considerably, especially on the head, shoulders and back. The darkest sables tend to resemble the lighter grades of the saddle pattern, particularly if the lateral V markings lack prominence. The typical sable is that of the Rough Collie or Shetland Sheepdog. Only A^y yellow is converted to sable by the umbrous polygenes. Although not definitely proven, these polygenes cannot produce a similar effect upon *e* yellow. Gene *e* is apparently epistatic to umbrous modification.

White Dogs

A completely white phenotype may be the basis for several breeds or be a variety of a breed. In principle, all-white dogs can be produced by several genotypes. The most common form in many mammals, but not the dog, is the albino. This phenotype is completely devoid of pigment, not only from the coat but also from the eyes. The outcome is a pink-eyed white. However, when breeders speak of a white individual they mean a black-eyed white. These are not albinos in any shape or form, for the albino always has pink eyes. Not even is the blue-eyed white a true albino. It is wise to be aware of the distinction.

One of the earliest suggestions for all-white dogs is that these are examples of extreme white spotting. This concept is formalized in Little's (1957) postulation of an extreme-white allele s^w. Animals of genotype $s^w s^w$ are either all-white or white, but for spots or patches in the region of the ears and at the base of the tail. Persistent selection in the past has meant that such individuals now constitute only a small proportion of the pups. However, the fact that such individuals are produced is a good guide to the $s^w s^w$ genotype and distinguishes it from other explanations. By this means, Little has been able to demonstrate that the white phenotype of the Bull Terrier, Great Pyrenees, Samoyed and Sealyham Terrier is $s^w s^w$.

Another explanation for a white phenotype is that of a chinchilla gene which removes all pigment from the coat but not from the eye. The result would be a black-eyed white. Tjebbes and Wriedt (1927) suggested that the Samoyed may be such an animal, but it seems that they were mistaken in view of Little's subsequent findings. However, this does not necessarily mean that all Samoyeds are $s^w s^w$. Little's data are drawn from American breeders, while Tjebbes and Wriedt's observations relate to Scandinavian dogs. On the other hand, Tjebbes and Wriedt's suggestion is at best probably only partly correct.

There are only two sorts of pigment in the pigment granules of the hair, the black (eumelanin) or yellow (phaeomelanin). If mutant genes could be discovered which remove both of these pigments from the coat, the result would be white. One does not have to look far to find genes removing black melanin, since two have been identified, A^y and *e*. These genes do not remove black physically, of course, they operate by causing the pigment granules to be yellow instead of black. This is a step in the right direction, however, for if an additional gene can be found removing yellow, the result would be a white coat. Such a gene would be the chinchilla c^{ch}, since the gene acts primarily by suppressing yellow pigment.

Although the existence of c^{ch} has been postulated by Little (1957), the evidence for it is not unambiguous. However, it is convenient to consider that c^{ch} does exist for the moment. The action of c^{ch} would be to sharply reduce the amount of yellow but leave black largely unaffected. Combining A^y with c^{ch} produces a white animal, because A^y removes all black and c^{ch} removes the remaining yellow. Since A^y does not remove black fully in all cases, it is possible to obtain sable shaded white or if the animal possesses a black mask a so-called "pointed white". The *e* is more effective in removing black pigment, hence the clearest whites would be $c^{ch}c^{ch}ee$ rather than $A^y{-}c^{ch}c^{ch}$.

However, the c^{ch} may not be fully effective in removing all yellow pigment and a residual may be left. Such animals would be similar to the pale cream Golden Retriever. Additional selection would be required to remove the lingering traces of yellow. There is little doubt that the intensity of yellow varies considerably in fully coloured (CC) animals, independently of an assumed c^{ch} allele. The variation is such that the necessity of posturing a major yellow diluting gene may be called into question. The lightening and virtual disappearance of yellow pigment may be due entirely to the action of minus rufus polygenes. White animals of the present category may be distinguished if they produce dogs with an overall cream tinge, especially along the spine, or are bred themselves from pale cream parents. Little (1957) considers the West Highland White Terrier to be an example of the A^y-$c^{ch}c^{ch}$ type of white.

The homozygous merle (MM) is white or almost white, with patches of coloured hair on the head. Some individuals develop into fine dogs, but the majority suffer from one or more defects. These include blue iris colour, deafness in one or both ears, and reduction in size of the eye, even to the point that one or both eyes may be absent. In general, it is advisable that white merle animals should not be bred.

Iljin (1932) states that the white variety of the Russian Shepherd Dog and of the Siberian is due to a dominant gene *W*. However, Whitney (1974) commenting on the white variety of the Siberian, states that, in his experience, it is inherited as a recessive. That a completely white coat can be inherited as a recessive is shown by the ample breeding data of Carver (1984).

Eye Colour

Eye colour in ordinary terms means iris colour, since the pupil appears black in most dogs. Normally the iris is brown, but the depth of colour may vary. The colour may be scored as dark, medium or light; or by such terms as brown, hazel or amber. Iris colour is determined by two factors, the effect of major genes, which is an incidental aspect of their influence on pigmentation, and polygenes, which modify the colour independently of the major genes.

Both the chocolate *b* and dilute *d* genes have an affect on iris colour. The irises of *bb* animals are regularly lighter than *B*–, usually being referred to as hazel, although the actual colour may vary. Dilute or blue *dd* also show lighter iris colour, being amber or even yellow, but again with variation. Burns (1943) describes the irises of blue animals as smoky or smoky-yellow, varying in depth to medium brown.

Burns (1943) observed that iris colour in normal dogs varied from dark to light brown and, although definite evidence was not obtained, concluded that the light iris colour was recessive to the darker tones Whitney (1974) considers that dogs with light brown irises mated *inter se* never produce pups with dark irises. Conversely, matings between dogs with dark irises can produce pups with light irises. Whether or not it is legitimate to postulate a major gene for dark versus light iris is conjecture at present. It is more probable that iris colour is determined by polygenes increasing or decreasing the amount of pigment in the iris and thus modifying its appearance. The same polygenes will be modifying the iris colour of chocolate and blue dogs, independently of the effects of *b* and *d* genes, and engendering the range of iris colour shown by these dogs.

The iris colour variation considered here is that exhibited by the ordinary run of dogs. There are special cases, however, which may be mentioned for completeness. Certain extreme albinotic alleles may have pale blue or white irises, due to severe limitation of the amount of pigment in the eye. If there is little, the eye is bluish or bluish-red, the colour being a mixture of pigment and the redness of blood showing through the translucent iris. If pigment is absent (as in the complete albino), the iris colour is white and the pupil is red. The pink-eyed mutant *p* produces bluish-red or reddish eyes due to some cause, a severe reduction in eye pigmentation.

The merle gene M may produce a blue iris as one of its affects. The iris may be partially or wholly blue, and one or both eyes may be affected. Again the cause is a reduction of normal eye pigmentation, in keeping with the diluting influence of M upon coat colour. The effect of M upon eye colour is as erratic as its effect on coat colour.

Iljin (1932) mentions a blue iris, which is independent of coat colour, as being inherited as a recessive to normal brown eye. Whitney (1947) cites a similar instance in a strain of Poodles. He states that the blue iris occurred in both black and liver-coloured individuals. No data is given, but Whitney opines that the aberrant irises was inherited as a recessive. The condition was treated as an anomaly and eliminated by selective breeding. Iljin (1932) also mentions a condition in which the normal yellowish-green reflection, seen when a beam of light is directed upon the eye in darkness, is replaced by red. The condition is said to be inherited as a recessive to normal reflection.

Nose Leather Colour

In contradiction to iris colour, the colour of the nose leather is almost wholly determined by major genes. The normal dog in good health has black nose leather. On the other hand, chocolate (*bb*) animals have a liver-coloured nose, while dilute (*dd*) animals have a slaty blue nose. There may be variation of depth of colour, but in general the nose colour is consistent with the colour of the coat. Even when the coat is red or yellow, *B*– animals can be distinguished from *bb* by the colour of the nose leather.

Coat Quality

There are a considerable number of different coat types in the dog. However, the differences may be more apparent than real. That is, the number of major genes may be fewer than one might expect, the variation of expression of these stemming from polygenic modification. The example of long hair will illustrate the situation. There is no reason to think that long hair is produced by other than one long-haired gene. Yet is is obvious that long-haired animals differ from breed to breed. This variation is due to polygenes modifying the expression of the long-haired gene to fit the conception of the breeders as to how the breed should appear. The diversity of appearance is due to the long period of selection by which man has shaped the breeds.

The correct method of analysing the coat quality would be to isolate the various components and to study the inheritance of these. Very little has been accomplished along these lines. Even in the normal coated dog (those without major mutants) there is a surprising variation in hair length, density (in the form of number of hairs per unit of skin area), number of guard hairs relative to number of underfur hairs, and straight versus a tendency to wavy coat. The above items cover the main components, but probably not all. Many of the components vary independently of each other, and between them can generate most of the coat types. The inheritance may be presumed to be polygenic until proved to be otherwise.

Long Hair

Long hair is due to a gene *l*, a mutant which probably prolongs hair growth and results in a coat several times the normal length. The expression has been modified in several respects. In some, the coat does not appear to be very long and lies flat against the body, the presence of *l* being indicated by feathering on the legs and tail, as in the English Setter. In others, it is especially long, straight and of silky texture, as in the Yorkshire Terrier. In others, it is full and very dense, standing out well from the body, as in the Chow Chow. Finally, in others, the coat is long, full and of a shaggy nature, as in the Old English Sheepdog. In the Spaniel, the long coat often has a wavy texture, and in the American Cocker Spaniel the coat is very developed on the legs. Feathering is discussed later.

Wire Hair

Wire hair, as found in the Airdale Terrier, Dachshund, Fox Terrier, Griffon Bruxellois and Pointer, is due to a dominant gene *Wh* (Whitney, 1947; Winge, 1950). It is assumed that the same gene is concerned in all of these breeds and by extension to all other wire-haired breeds. It may be objected that this could be unwise. However, justification may be found in the scientific principle of economy of hypothesis, which says that the minimum of assumptions are to be made to explain a given situation. In the present case this is that wire hair in all breeds is due to the same gene. At this moment there is no means of knowing whether there is more than one wire-hair type of mutant in the dog, although this is assumed as a working hypothesis. As yet unproven, it may be that *Wh* tends to inhibit hair growth, so that *llWh–* has a longer coat than *LLWh–* but not as long as *llwhwh*.

Kinky Coat

Whitney (1947) has contributed evidence that the curly or kinky coat of the Irish Water Spaniel is inherited as a recessive to straight coat. The data are scanty but suggestive. The gene is provisionally symbolized as *k*.

Curly Coat

It might be supposed that the characteristic coat of the Curly Coated Retriever is a short-haired version of the curly coat of the Irish Water Spaniel. However, Burns and Fraser (1966) cite a report that a litter from a mating of a Curly Coated Retriever to a Pointer had curly coats like the Retriever. This report is at variance with the observations of Whitney given above. It could be a salutary warning not to assume that all curly or wire-haired characteristics are due to the same gene.

Wavy Coat

Whitney (1947) has given details of crosses which indicate that a wavy coat, such as shown by the Cocker Spaniel and a few other breeds, might be inherited as a recessive to straight hair. However, this is little more than an educated guess, and the postulated gene may only be tentatively symbolized as *wa*. The maximum expression of waviness is found in long-haired breeds.

Hair Whorls

Exceptional whorls of hair in a strain of Cocker Spaniels have been reported by Pullig (1950). These occurred mainly on both shoulders or on both flanks, but some dogs had only one whorl; or the whorl may be on the head or neck. The condition was inherited as a recessive, due to a gene *wo*.

Ripple Coat

Whitney (1947) has described a curious anomaly either of skin folds or hair direction of young pups in a strain of Bloodhounds. Between the ages of 1–7 days the affected pup shows long regular waves or ripples extending from the neck to the hind legs, spreading slightly as they travel posteriorly. After 7 days of age, the ripples disappear and the coat appears normal. The breeding data are suggestive of the inheritance of a recessive gene *rp*.

Feathering

Feathering, or a fringe of longer hair, on the inside of the legs and undersurface of the tail is a feature of long-haired dogs. However, the extent of the feathering can vary from slight to extensive. Whitner (1947) commented that crosses between Irish Water Spaniels (long-haired but little tail feathering) and either Cocker Spaniels or Irish Setters (both long-haired breeds with ample tail feathering) gave offspring which were long-haired with tail feathering. These results indicate dominance of tail feathering, but are inadequate to show if the presence or absence of tail feathering is due to a single gene or if it is a polygenic variation of long hair. The latter is the most likely explanation, and the character may be cited as an example of how the expression of the long-haired gene can be modified by other genes.

Hair Growth and Colour

Whitney (1947) makes the point that in long-haired white-spotted dogs the hair growth may be different between the pigment and white areas. He says that in the Cocker Spaniel and English Setter hair growth on black spots appeared to be longer than that of neighbouring white hair. Whitney also states that black hair grows longer than red hair. Burns and Fraser (1966) evidently agree with this observation, and add that while white hair may grow as long as the black, it tends to be softer in texture. Yellow or red hair tends to be finer and more silky than black.

Hairlessness

Hairlessness is strictly an anomaly. However, the condition may be mentioned here because it has been claimed as the hallmark of breeds of dog, e.g. the Chinese Crested and Mexican Hairless. These animals are often not completely hairless, but have variable amounts on the head and feet. Letard (1930) showed that one strain of hairless dogs was due to a dominant gene *Hr*, and that the homozygote was lethal. That is, all of the living hairless animals were heterozogotes *Hrhr*. The conclusion of dominant heredity with lethality of homozygotes has been confirmed for the Chinese Crested dog by Robinson (1985). Not all cases of hairlessness are produced by dominant genes, for possible recessive form in Spanish dogs has been described by Zulueta (1949). More recently, Sponenberg *et al.* (1988) have shown that the hairless condition of the American hairless terrier is due to a recessive gene *ha*.

Hairless mutants recur from time to time and some may survive in the warmer regions of the world or are preserved by man as a curiosity. If they are of some antiquity, these animals can acquire an aura of respectability and be recognized as a breed. Whether or not this is justifiable can be a matter of opinion.

Comparative Gene Symbols

Those readers who read widely in genetics will soon encounter the problem of confusing symbolism. Not that of the genes themselves, nor that of the practical aspect of remembering. No, no. The problem is that different authors have employed different symbols for the same genes. This is regrettable but perhaps unavoidable, because, unlike many other mammals, a standard symbolism has not been agreed between canine geneticists. This section is in effect a plea for a standard nomenclature.

The first thoroughgoing discussion in English of dog genetics is that of Dawson (1937), who made a serious attempt to present a distillation of previous investigations. Dawson's compilation has withstood the test of time quite well. Both Little's (1957) and Burns and Fraser's (1966) subsequent contributions are in the mainstream of mammalian genetic thought. Alas, the same cannot be said for that of Winge (1950). Winge's system represents a complete break with precedence and is apt to be confusing. Several genes are proposed for which there is no need. Several others are placed incorrectly in the allelic series. Two symbols given in an initial list of definitions are not used in the same sense in the subsequent text.

TABLE 11

Coat colour and texture mutant genes, their symbols and names

Symbols in brackets indicate provisionally recognized genes.

Symbol	Designation	Symbol	Designation
A^s	Solid black	Hr	Hairless
A^y	Dominant yellow	k	Kinky coat
A	Agouti	l	Long hair
a^{sa}	Saddle	M	Merle
a^t	Black and tan	(Ma)	Mask
a	Non-agouti	s^i	Irish spotting
b	Brown	s^p	Piebald spotting
c^{ch}	Chinchilla	s^w	Extreme white
c^b	Blue-eyed albino	p	Pink-eyed dilution
c	Albino	rp	Ripple coat
d	Dilute	T	Ticking
E^{br}	Brindle	Tw	Tweed
e	Non-extension	w	White
f	Absence of flecking	(wa)	Waved coat
G	Progressive greying	Wh	Wire hair
H	Harlequin	wo	Whorls
ha	Hairless, American		

TABLE 12

Check List of Comparable Symbols Adopted by Various Authors

This book	Little (1957)	Burns and Fraser (1966)	Dawson (1937)	Winge (1950)
A^s	A^s	A	E^d	C^l
A^y	a^y	a^y	A^y	A
A	a^w	a^g	A	C
a^{sa}	—	a^s	—	c^{sa}
a^t	a^t	a^t	a^t	c^{bi}
B	B	B	B	B
b	b	b	b	b
C	C	C	C	F
c^{ch}	c^{ch}	c^d	c^r	f
c^b	—	c^b	c^b	—
c	c^a	c	—	—
cn	—	—	—	—
D	D	D	D	D
d	d	d	d	d
E^{br}	e^{br}	e^b	e^p	c^{br}
E	E	E	E	E
e	e	e	e	e
G	G	—	—	—
l	—	—	—	k
M	M	M	V	H
Ma	E^m	E^m	—	c^{ma}
p	p	—	—	—
pp	—	—	—	—
S	S	S	S	T
s^i	s^i	s^i	—	—
s^p	s^p	s^p	s	t
s^w	s^w	s^w	w	—
Sg	—	—	—	—
T	T	—	T	S
Wh	—	—	—	W

7.

Genetics of Breeds

It is almost impossible to present a detailed discussion of the genetics of all breeds of dogs. In the first place, there are simply too many breeds to cover adequately. In the second, there is a total lack of information for many breeds. This does not mean that the situation is hopeless. A fair amount of information has been steadily accumulated over the last 60-70 years and a number of breeds have been studied in detail. A point of interest is that many breeds bear a close resemblance to each other, implying either remote common ancestry or parallel evolution by breeders selecting for a similar type of dog. Even the coat colour and type featured in this selection. Looking at the situation from another angle, it is evident that the number of different colour genes are relatively few, and information gathered for one particular breed is valid for breeds similar to it and often for breeds less closely related. This widens the scope of inquiry, and means that sensible conclusions can be made for a sizeable number of breeds.

Most of the discussion of this chapter will be centred upon the genetics of coat colour and type. This is inevitable, because these characters are primarily governed by major genes, regardless of how the expression of these genes may vary from breed to breed. This means that the pertinent genetic formulae can be written for the colour varieties making up different breeds. Colour variation is not the only attribute of a breed, in fact arguably not the most important. Two important items are the conformation and stance, but these attributes are controlled by polygenic variation which cannot be expressed as symbols. Characters of this nature can only be manipulated by selective breeding, by careful choice of breeding stock which conforms closely to the breeder's ideal. The relevant breeding techniques to achieve this aim have been outlined in a previous chapter. In short, these are the drafting of a total score index to facilitate selection programmes, coupled with mild inbreeding.

Three books have discussed the genetics of colour inheritance in specific breeds, namely Whitney (1947), Winge (1950) and Little (1957). Whitney's book summarized most of the previous studies, with additional observations of his own. Winge's account is interesting but unwarrantedly marred by an idiosyncratic symbolism of the mutant genes. Little's contribution is the most extensive and is outstanding because it is based upon the valuable work performed at the Jackson Laboratory in the USA and upon an analysis of breeder's records. To these may be added Willis's (1976) monograph on the German Shepherd dog.

It is possible to discuss the colour genetics of some 115 breeds. Fortunately, these include many of the more popular breeds. This is to be expected, of course, for the greater the number of animals bred, the greater are the chances of discovering useful information. Moreover, those breeds which have been considered are reasonably representative of dog breeds as a whole. The other side of the coin is that many less common breeds cannot be considered because of lack of reliable information. If your favourite breed has not been discussed, please refer to one or more breeds which are similar to it. In all likelihood, it will be found that the genetics are identical or, if not identical, sufficiently close that intelligent guesswork can close the gap. Unfortunately, guesswork, however intelligently applied, is no substitute for data. An effort should be made to seek out reliable information for analysis. If need be consult a geneticist to ensure that the correct information is collected and presented.

In the discussion of breed genetics certain conventions have been observed. With certain exceptions, which should be obvious from the context, only mutant genes will be depicted in the genotypes. It may be assumed that genes at other loci will be non-mutants. This simplifies the genotypes so that the eye can grasp the essentials. In the genotypes, a dash (–) after a gene means that the gene may be present in a homozygous or heterozygous condition. This applies to dominant genes, and it is often unwise to assume that these are present as homozygotes. No other implication should be read into the provision of dashes.

A few words of warning may be in order. It is unfortunate that breeders tend to be confusing with descriptions or terms for the various colours and patterns. Different people may use the same term for dissimilar colours or different terms for the same colour. Three common errors are the misinterpretation of liver as red, the confusion of saddle pattern with black and tan, and confusion of saddle pattern as sable or vice versa. There are probably others with which readers may have had personal experience or can call to mind. In the case of black and tan, the confusion has even reached into the wording of standards, where a reference to black and tan may actually mean saddle pattern.

One likely source of error is that of distinguishing between reds and yellows due to either A^y or e genes. The majority of reds are A^y, and these may be distinguished by the presence of dark ticking and/or the presence of a mask. There is reason to think that e reds never or rarely show dark ticking and never exhibit a mask. Unfortunately, it is unsafe to assume that all clear reds or yellows are *ee*. Also, it is unwise to assume that all reds are A^y and all yellows are *ee*. Where there is doubt of the nature of red or yellow, it may be assumed initially to be A^y. As a point of policy, anyone who supposes that red or yellow is *ee* should feel under an obligation to show that it is not A^y.

Where white spotting is an integral part of the make-up of a breed, this is discussed in terms of the three alleles postulated by Little (1957). No account will be taken of minor white spotting which occurs in many breeds. In the majority of cases, this spotting is referred to as a fault. Minor white spotting is the occurrence of white spots on the nose, breast, stomach and tail tip, rarely on the paws. The frosting on the chin of many breeds is not minor spotting. It is doubtful if minor white spotting is due to a single gene. If it is, the gene often fails to express itself. The spotting is probably polygenic with variable expression. Persistent selection against animals showing the spotting is the only means of bringing it under control. In some breeds, minor spotting is listed as a fault, but not a serious one. Such a relaxed attitude is understandable, but it throws the onus of deciding whether or not to cull upon the breeder. It is not the ideal method of eliminating a fault.

The genetic designation of coat quality presents quite a problem. Only a few genes have been definitely established, long hair and wire hair being the most defined. Yet many different coats exist. Many of these are probably different expressions, as modified over the years as a breed characteristic, of the same gene. Long hair is probably a good example of the process. Only one long-haired gene probably exists, but its expression differs in many breeds (Whitney, 1947).

One cannot say the same for curly, kinky, rough or wire-haired animals. Several different mutant genes may be involved in different breeds to produce these various coats. In the domestic cat, for example, several mutants for the rex coat have been identified, but phenotypically they are very similar. It is easy to visualize the problem. The same gene could have a somewhat different expression in different breeds, due to the breed genotype, and be confused with different genes producing different phenotypes. On the other hand, it is unwise to make claims for new genes unless the evidence is reasonably sound. It is a principle of genetics that only the minimum of genes are postulated until the evidence decrees otherwise.

In undertaking a reasonably comprehensive discussion of breed genotypes, it is easy to make false assumptions for the presence or absence of certain genes. It is also easy to make straightforward errors. Notification of probable mistakes would be appreciated. Also, any information which would enable additional breeds to be sensibly discussed would be welcomed.

Affenpincher

In the main, this is a black dog with a long wiry coat. Black and tan, as well as sables of varying shades, are also recognized. According to Little (1957), black masks are present in some individuals.

Colour	Genotype
Black	A^s–*llWh*–
Black and tan	a^ta^t*llWh*–
Sable	A^y–*llWh*–

Afghan Hound

The Afghan Hound is bred in a number of colours. Blacks are known, as also black and tan, indicating the presence of genes A^s and a^t. Reds are also common, many with masks and sable shading, and this indicates the presence of A^y. The breeding of black and tan pups from matings of two reds (Little, 1957) is confirmation if such is required. Little also makes reference to blue animals, but says that these are rare. He also states that breeding experiments at the Jackson Laboratory have revealed that some reds are *ee*. The long silky coat indicates that the Afghan is *ll*.

In addition to red, yellow, cream and almost white animals are known. These are beautiful and striking dogs, especially when combined with a black mask. Little postulates that the c^{ch} gene is responsible for some of the pale colours in conjunction with A^y and *e*, the c^{ch} having a diluting effect on yellow but not on black pigment. However, it is equally likely that rufus polygenes are involved, the plus polygenes producing the red and the minus polygenes the yellow and creams. The long hair is contributory to this effect, by exposing the paler undercolour and producing a paler colour than would appear on a short-haired dog. Some of the paler colours could be dilutes (*dd*), especially those without dark masks. In conjunction with c^{ch}. or minus rufus polygenes, such individuals could be exceptionally pale.

Colour	Genotype
Black	A^y–*D*–*ll*
Black and tan	a^ta^t*D*–*ll*
Red/yellow	A^y–*D*–*ll*
Blue	A^s*ddll*
Cream	A^y–*ddll*

Airedale Terrier

This breed represents the archetypal saddle pattern, with a well-defined V-shaped black area on each side of the body and extensive tan areas on each of the legs and face. Some light ticking occurs in the saddle, but as a rule this is minimal. The coat is wiry. Genotype: a^{sa}-Wh-.

Alaskan Malamute

The colour is basically wolf-grey but variable. In fact the variation is interesting in that it illustrates that the wolf-grey pattern can vary from light grey to a blackish grey. Some individuals have a characteristic light grey face while, at the other extreme, some have a black mask. Genotype: A- *or* a^{sa}-, if the two genes are equivalent.

American Cocker Spaniel

This breed is fundamentally the same as the English Cocker Spaniel, and details of the various colours may be found in the section on the latter. It is doubtful if any major change has occurred in the colour genetics during the evolution of the American version. The very full development of the long hair on the legs of the American breed, however, is indicative of the effects of modifying genes on the regional development of long hair. Both the American and English Cocker Spaniels are long-haired breeds, but the profusion of dense long hair on the stomach and legs of the former is noteworthy.

American Foxhound

The American Foxhound is a larger dog than the Foxhound (*q.v.*), but the coat colour genetics are identical. According to Little (1957), nonspotted animals are becoming frequent. Whitney (1947) writes of a merle Foxhound under the name of calico.

Colour	Genotypes
Yellow or tan	A^{y}–S–
Saddle	a^{sa}–S–
Yellow, white	A^{y}–s^{p}–
Saddle, white	a^{sa}–s^{p}–

American Hairless Terrier

The American Hairless terrier possesses a sparse covering of downy hair as a pup, which is lost as an adult. The hairlessness is produced by a recessive gene *ha*. Consequently, the American Hairless terrier breeds true for the hairlessness; this is in contrast to other hairless breeds which are due to the heterozygosity of a dominant gene (Sponenberg *et al.*, 1988).

American Staffordshire Terrier

This breed is a slightly larger version of the Staffordshire Bull Terrier (*q.v.*). The genetics of coat colour appear to be identical.

American Water Spaniel

Apart from its smaller size, this breed is identical to the Irish Water Spaniel and has the genotype A^s-*bbkkll*.

Anatolian Shepherd Dog

The usual colour varieties are the fawn, tricolour and white. The fawn is A^y with a black mask of varying intensity. Additionally, most fawns exhibit variable amounts of sable shading. The tricolour is fawn plus piebald spotting (s^p). The white has the genotype $A^y c^{ch}$ and may be pale cream in colour. The fawn varies in the intensity of yellow, from deep fawn to red. Both short and long coat occur in the breed (Robinson, 1989*a*).

Colour	Genotypes
Fawn	A^y
Tricolour	$A^y\text{–}s^p s^p$
White	$A^y\text{–}c^{ch}\text{–}$

Australian Silky Terrier

These are saddle pattern animals of a rather unusual grey-blue and tan colouring. The breed is similar to the Yorkshire Terrier (*q.v.*) and it is probable that an identical genetic situation exists.

Australian Terrier

Dogs of this breed are saddle pattern, but of unusual colouring. The blue and silver grey body colour is similar to that of the Yorkshire Terrier (*q.v.*) and it is probable that a similar genetic situation exists in the two breeds.

Basenji

The colour of this sprightly dog is usually light red, either clear or slightly ticked with black along the spine. This is indicative of A^y yellow. It would seem that tan pattern and possibly blacks have been known in the past. White spotting is regularly present, but of the restrictive Irish grade.

Colour	Genotypes
Black	$A^s\text{-}s^i\text{-}$
Black and tan	$a^t a^t\text{-}s^i$
Red	$A^y\text{-}s^i\text{-}$

Basset Hound

The colour of the Basset appears to be either yellow or black and tan, the latter probably being saddle pattern in actuality. All dogs are white spotted and of the piebald grade of markings. Some ticking is evident, especially around the muzzle. If this is due to the T gene, the expression is notably restricted.

Colour	Genotypes
Yellow, white	$A^y\text{-}s^p\text{-}$
"Black and tan", white	$a^{sa}\text{-}s^p\text{-}$

Beagle

The Beagle is another of the archetypal saddle pattern of the a^{sa} gene. The pattern is relatively stable, although, if a large number of Beagles are examined, some variation of the extent (and to a lesser extent the intensity) of the saddle may be observed. The intensity of the reddish areas of the pattern also varies to some extent. All of this variation may be attributed to polygenic variation. White spotting of the "collar" (usually) or piebald grades is invariably present. The genotype is thus $a^{sa}\text{-}s^i\text{-}$ or $a^{sa}\text{-}s^p\text{-}$. Little (1957) mentions that ticking T may be seen in some individuals.

It is interesting to read that several other phenotypes are described as rare segregants in 160 litters reported by breeders to the Jackson Laboratory (Little, 1957). Liver, blue, yellow and a particular dull yellow in the normal red areas (saddled unaffected) expected of chinchilla were listed. These pups indicate the presence of b, d, e, and possibly c^{ch}, respectively. However, none of these was common, representing merely a few individuals per 100 pups, and in no way distracting from the character of the usual Beagle. Their low frequency means that they are not typical of the breed as a whole and probably owe their appearance to a long-forgotten outcross or mutation. Beagles, as a breed, are probably not unique in being predominantly of one colour, but have other colours as rarities.

Bedlington Terrier

According to Little (1957), the pale coloration typical of this breed is due to the dominant greying gene G. In the case of the blue, the pups are born black but change gradually to light blue with age. Assuming that G is present in all of the varieties, examination of the colour of the pups is the only sure means of discovering their genetic nature. To attempt a similar classification with the adult coloration is to court mistakes.

In discussing the breed's genetics, Little makes several dubious statements. Presumably, the A^s gene is responsible for the solid colours; however, because some of these show lighter pigmentation where the tan of the tan pattern is located, he concludes that the a^t is involved. This is not necessarily so. The fact that some blues are very light and some sandy animals become cream in colour prompts him to conclude that the c^{ch} gene is involved. Since c^{ch} has little effect on blue, and if it is necessary to invoke a second diluting gene, it is more probable to be d. Little depicts sandy as being liver diluted by c^{ch} (or by d). This is possible, but it may be wise to keep an open mind that either A^y or e are involved. All of these points at issue could be resolved by noting the puppy colour before the greying sets in.

Colour	Genotypes
Blue	$A^s\text{–}B\text{–}D\text{–}G$
Blue and tan	$a^ta^tB\text{–}D\text{–}G\text{–}$
Liver	$A^s\text{–}bbD\text{–}G\text{–}$
Liver and tan	$a^ra^tbbD\text{–}G$
Sandy	$A^y\text{—}D\text{–}G\text{–}$
Cream	$A^y\text{—}ddG\text{–}$

The coat is peculiar. It is dense and of linty appearance, and no information seems to be available on its mode of inheritance. It would be a modified wire-hair (Wh), but this is only a tentative suggestion. For reasons discussed above, some of the genotypes could be incorrect.

Belgian Shepherd Dog

The breed is bred in four distinctive varieties, namely, the Groenendael, which is solid black with a long coat; the Tervueren, which is of sable coloration with a black mask and long coated; the Malinois, which is also of sable coloration with a black mask but short coated; and the Lakenois, which is also of sable coloration with a black mask but is wire haired.

The Groenendael has proved to be due to the dominant black allele A^s while the other three owe their colour to the dominant yellow allele A^y. The primary differences between the three reside in the coat type, the Tervueren and Lakenois possessing the long hair (l) and wire hair (Wh) genes, respectively. The Tervueren has two colour phases, the red and the grey, where the red pigment is degraded to a light yellow. The grey is inherited as a recessive to the red and is probably produced by the chinchilla allele c^{ch} (Robinson, 1988).

However, more recent observations have revealed that some of the greys are agouti (A), with a light blue undercolour at the base of the hairs, instead of pale cream or whitish undercolour. Recently, completely black pups have been bred from two Tervueren parents, which is suggestive that a small proportion of Groenendael could be genetically non-agouti (aa).

On occasion, dogs may be found which combine some of the attributes of the above varieties, such as a black Malinois or Lakenois. Some years ago, a "recessive black" was common in the breed but this turned out to be black and tan (a^t) with a black mask covering up the usual facial tan markings of the pattern.

Colour	Genotypes
Groenendael	A^s–C–$llWhWh$
Tervueren, red	A^y–C–$llWhWh$
Tervueren, grey	A^y–c^{ch}–$llWhWh$
Malinois	A^y–C–L–$WhWh$
Lakenois	A^y–C–L–$WhWh$

Bernese Mountain Dog

The colour is that of a black and tan, with white markings restricted (ideally) to the breast, paws and a forehead blaze. The coat is long. Genotype: $a^t a^t lls^i$-.

Black and Tan Coonhound

A typical black and tan of genotype $a^t a^t$.

Bloodhound

The usual colours are red and "black and tan", the latter being saddle pattern in genetic terms. Some reds may possess a mask. Liver and blue saddle animals are known but presumably uncommon.

Colour	Genotypes
Red	A^y—D–
"Black and tan"	a^{sa}–B–D–
"Liver and tan"	a^{sa}–bbD–
"Blue and tan"	a^{sa}–B–dd

Border Collie

Several colours are known in this breed, but the predominant variety is the black with restrictive white markings. Genotype: A^s-s^i-.

Borzoi

This breed is a sable with piebald markings and a long silky coat. The sable suffusion is extensive and tends to exclude the alternative possibility that the colour could be saddle pattern. Black and white animals are known, but are not favoured.

Colour	Genotypes
Black	A^s–lls^p–
Sable	A^y–lls^p–

Boston Terrier

Black and brindle are the predominant colours of this breed, both combined with spotting of the "collar" grade. The white pattern is an important feature of the breed and must be symmetrical. As Little (1957) points out, selection of Irish spotting for maximum expression is the best means of achieving this, because Irish spotting tends to produce a natural symmetric pattern, whereas piebald spotting does not. Little notes that almost white pups occur as a result of segregation of the s^w gene. This suggests that some typically white pattern animals could be Ss^w. Some blacks could be due to the A^s gene or be very dark brindles, without any trace of yellow hairs.

Colour	Genotypes
Black	A^s–s^i–
Brindle	A^y–E^{br}–s^i–

Boxer

The Boxer occurs in only two colours, fawn and brindle, the latter being variable, ranging from fawn-brindle to black-brindle in the amount of dark striping. A black mask occurs in almost every animal. Those with white markings display the restrictive Irish spotting. According to Little (1957), the all-white extreme-white grade of spotting ($s^w s^w$) occurred in the breed, but is now regarded with disfavour.

Colour	Genotypes
Brindle	A^yE^{br}–S–
Red, fawn	A^y–E–S–
Brindle, white	A^y–$E^{br}s^i$–
Red, fawn, white	A^y–E–s^i–

Brittany Spaniel

The permitted colours in the breed are liver and white and orange and white. The light or liver noses of the orange and white animals indicate that the breed is homozygous for *b*. The white markings are those expected for s^ps^p animals, the coloured areas appearing as larger patches. The white areas may be clear or be lightly ticked. Little (1957) considers that the orange colour is produced by the *e* gene. The coat is long.

Colour	Genotypes
Liver, white	bbE–lls^p–tt
Liver, white, ticked	bbE–lls^p–T–
Orange, white	$bbeells^p$–tt
Orange, white, ticked	$bbeells^p$–T–

Bull Terrier

This breed is differentiated into coloured and white varieties and there is dispute whether or not they should be interbred. For convenience, the colours will be considered first. The usual colours are black, brindle and red or deep fawn. The latter variation is due to rufuls polygenes. The reds and fawns may show various degrees of mask pigmentation on the face. Many animals show white spotting of the "collar" grade and are probably s^is^i. Briggs and Kaliss (1942) remark that black and tan (a^ta^t) and blue (A^s-dd) occurred in the breed in the past.

The white variety is s^ws^w. This is revealed primarily by the presence of pigmented patches on the heads of a small proportion of dogs (Briggs, 1940: Little, 1957). The genes carried by the white animals may be any of those discussed above, but, of course, masked by the white fur. Some of the coloured patches may be large enough for the "hidden" genes to be recognized. The white Bull Terrier will breed true, even if bred initially from coloured parents, a point stressed by Briggs (1940) in a paper written to reassure breeders on this point.

Colour	Genotypes
Black	A^s–E–s^i–
Fawn	A^y–E–s^i–
Brindle	A^y–E^{br}–s^i

Bulldog

Red or sables of various shades are common in this breed, together with brindle. Each of the above may have or not have a black mask. The fawny colour known as "fallow" may be red diluted by a chinchilla gene or by minus rufus polygenes. White spotting of all grades may be present and the variation is so wide that all three alleles, s^i,s^p and s^w, may be depicted. The white areas of these may be ticked by the *T* gene.

Colour	Genotypes
Red	A^y–E–S–
Sable	A^y–E–S–
Brindle	A^y–E^{br}–S–
Red, white	A^y–E–s^p
Sable, white	A^y–E–s^p
Brindle, white	A^y–$E^{br}s^p$–

Bullmastiff

The genetics of coat colours in this breed is exactly as those for the Boxer, except that white markings are discouraged. The depth of yellow pigment varies from reddish fawn under the influence of rufus polygenes. A black mask is prominent, but Little (1957) notes that few animals have the mask in the United States.

Colour	Genotypes
Brindle	A^y–E^{br}–
Red, fawn	A^y–E–

Cairn Terrier

This is another difficult breed to analyse, as pointed out by Little (1957). He considers the breed to be fundamentally a brindle (E^{br}) in which the brindling is confused or well broken up. The thick shaggy coat would aid the confused appearance. The colour varies from blackish to reddish, with many intermediate colours, according to the proportions of black and yellow hairs. The intensity of the yellow is variable, causing some animals to have sandy "undercolour" while others are cream or almost white, such as may be governed by variation in the rufus polygenes. The contrast between the dark head and the body could be due in part to the long hair of the body and the shorter hair on the head and to the presence of a black mask.

The red, sandy and cream varieties are probably A^y, with variable expression due to rufus polygenes. It is possible that some of the colours, discussed above as brindle, could be sable.

Colour	Genotypes
Brindle	$A^y\text{–}E^{br}\text{–}ll$
Red, sandy	$A^y\text{–}E\text{–}ll$

Cavalier King Charles Spaniel

The colour varieties of the breed are clearly defined. These are the red (known as ruby) and the black and tan. Workman and Robinson (1990) have demonstrated that ruby has the genotype *ee*. Both of the above colours may be combined with piebald spotting to produce the Tricolour and Blenheim varieties, respectively. The coat is long and silky in texture.

Colour	Genotypes
Black and tan	$a^ta^tE\text{–}llS\text{–}$
Tricolour	$a^ta^tE\text{–}lls^p\text{–}$
Ruby	$\text{—}eellS\text{–}$
Blenheim	$\text{—}eells^p\text{–}$

Chesapeake Bay Retriever

Animals of this breed are solid brown of varying shades, the standard allowing for variation from dark brown to faded tan or "dead grass". The fundamental genotype is thus $A^s\text{–}bb$. Little (1957) has proposed that the lighter colours could be due to the presence of a chinchilla type gene. This could be the case, for animals of $A^s\text{–}bbc^{ch}c^{ch}$ genotype would be expected to be of a khaki-like colour, judging from the effects of c^{ch} on brown colours in other species. However, as Little acknowledges, his proposal is speculative in the absence of experimental data for the definite existence of c^{ch}.

The coat has a tendency to be marcel waved, but there is insufficient evidence on the mode of inheritance of this character other than to say that it is inherited. The degree of waving can be controlled by selective breeding, but whether a major gene with variable expression is involved is unknown.

Chihuahua

These dainty little dogs are bred in a wide variety of colours. Black, black and tan and sables occur, as do liver, liver and tan and liver sables. Little (1957) notes that blacks have been produced from matings of two reds, a result which indicates that *e* yellow may also be present in the breed. However, the extent is unknown. The reds and sables vary in intensity, so much so that either a c^{ch} gene exists or the minus rufus polygenes are so numerous as to produce almost white animals. Blue individuals have been seen but are rather rare. White spotting is very common, both Irish and piebald forms. Little has indicated that extreme-white spotting may occur. All of the above colours are bred in smooth-coated and long-haired versions.

All of the following genotypes for the main colours may be combined with white spotting by substituting the relevant allele (s^i s^p *or* s^w for *S* and/or long hair by substituting *l* for *L*.

Colour	Genotypes
Black	A^s–B–D–L–S–
Black and tan	$a^t a^t B$–D–L–S–
Liver	A^s–bbD–L–S–
Liver and tan	$a^t a^t bbD$–L–S–
Blue	A^s–B–ddL–S–
Blue and tan	$a^t a^t B$–ddL–S–
Sable	A^y–B–D–L–S–
Liver sable	A^y–bbD–L–S–
Red (black nose)	A^y–B–D–L–S–
Red (liver nose)	A^y–bbD–L–S–

Chinese Crested Dog

This breed may be taken as a typical example of the hairless dog. The animal is not completely hairless, for variable amounts of hair occur on the head, lower parts of the legs and tail. The Mexican Hairless dog is virtually hairless, while the Chinese Crested is well endowed on the extremities. The reason is that the Mexican Hairless has a short coat while the Chinese Crested has a long coat. This is proved for the latter by the long-coated dogs, known as powderpuffs, which are produced by Crested matings. The breeding of a proportion of powderpuffs cannot be avoided because all Crested are necessarily heterozygous *Hrhr* (Robinson, 1985). The skin of the body may be plain or spotted, the skin of the latter being dark with piebald markings. The light piebald areas are usually dotted with dark spots. It is difficult to speculate which colour genes are present because of the absence of body hair.

Colour	Genotypes
Plain	$Hr\text{–}llS\text{–}$
Spotted	$Hr\text{–}lls^p\text{–}T\text{–}$

Chow Chow

Black, blue and various shades of red are known in the breed. Blacks are due to the A^s while the blue is due to *d*. According to Little (1957), reds may be either A^y or *e*. Some of the paler colours, fawn or cream, could be due to *d* or minus polygenes of the rufus group. The coat is long and exceedingly well developed all over the body.

Colour	Genotypes
Black	$A^s\text{–}D\text{–}ll$
Blue	$A^s\text{–}ddll$
Red	$A^y\text{–}D\text{–}ll$
Cream	$A^y\text{–}ddll$

Clumber Spaniel

The amount of white is extensive in this Spaniel, suggesting the presence of s^w instead of s^p. The coloured patches usually occur on the head and are varying shades of orange or yellow. Little (1957) believes that the colour is due to the *e* gene. The nose leather is usually liver or flesh colour, suggestive of the presence of *b*. This genotype is $bbeells^ws^w$.

Coarse-haired Griffon

This is a liver or chestnut coloured dog, with a solid coloured head and roan body. The coat is long and coarse to the touch. Its genotypes is probably $A^s\text{–}bblls^p\text{–}T\text{–}Wh\text{–}$ *or* $A^s\text{–}bblls^ws^wT\text{–}Wh\text{–}$. The roan effect is created, or enhanced, by the long wiry coat, producing a mixture of white and dark coloured hairs instead of the more usual ticking associated with *T*.

Curly-coated Retriever

As regards colour, animals of the breed are either black or liver, having the genotypes $A^s\text{–}$ and $A^s\text{–}bb$, respectively. However, the primary characteristic is the presence of small tight curls all over the body. These are probably due to the same recessive gene as found in the Irish Water Spaniel. If this is so, the genotypes may be written as: black $A^s\text{–}B\text{–}kk$ and liver

A^s–*bbkk*. However, Burns and Fraser (1966) have commented that a single litter of pups from a mating between a Curly-coated Retriever and a Pointer had curly coats. This implies a dominant mode of heredity of the kinky coat.

Dachshund

An extraordinary variety of colours and coats may be found in the Dachshund, as well as breed differences of body size. The majority of reds are almost certainly A^y, as shown by the presence of a faint mask in some individuals and a sable variety. Black and tan and liver and tan are common colours. The curious dapple variation is due to the heterozygosity of the *M* gene. An unusual colour is a wolf-grey or grizzled which could be either wild type (*A*) or light saddle (a^{sa}). As regards coat type, there are normal or smooth, long-haired and wire-haired; the last two being due to the *l* and *Wh* genes, respectively.

A representative list of varieties and their genotypes is as follows:

Colour	Genotype
Black and tan	$a^t a^t B\text{–}L\text{–}mmwhwh$
Liver and tan	$a^t a^t bbL\text{–}mmwhwh$
Red or sable (black nose)	$A^y\text{–}B\text{–}L\text{–}mmwhwh$
Red or sable (liver nose)	$A^s\text{–}bbL\text{–}mmwhwh$
Dappled (black, bluish)	$a^t a^t B\text{–}L\text{–}Mmwhwh$
Dappled (liver, dove)	$a^t a^t bbL\text{–}Mmwhwh$
Dappled (red, yellow)	A^y——$L\text{–}Mmwhwh$
Black and tan, long-haired	$a^t a^t B\text{–}llmmwhwh$
Liver and tan, long-haired	$a^t a^t bbllmmwhwh$
Red, long-haired	$A^y B\text{–}llmmwhwh$
Dappled, long-haired	$a^t a^t B\text{–}llMmwhwh$
Black and tan, wire-haired	$a^t a^t B\text{–}L\text{–}mmWh\text{–}$
Liver and tan, wire-haired	$a^t a^t bbL\text{–}mmWh\text{–}$
Red, wire-haired	A^y——$L\text{–}mmWh\text{–}$
Dappled, wire-haired	$a^t a^t B\text{–}L\text{–}MmWh\text{–}$

Dalmatian

This uniquely patterned dog is the outcome of countless generations of selective breeding. The foundations for an understanding of the genetics of the pattern were laid by Little (1957) when he showed that the genotype was $s^w s^w TT$. However, the coloured spots of this phenotype are usually interspersed with white hairs (flecking), whereas, those of the Dalmatian are solid. Schaible (1981) subsequently discovered that the absence of interspersed white hairs in the spots is due to a recessive gene *f*.

This is the story as far as the main genes for the pattern is concerned, while selective breeding has ensured that the spots are of even size and evenly distributed throughout the coat. The pattern is inherited independently of colour and may be black, chocolate or liver, and yellow. Little was able to show that the yellow is produced by the non-extension allele *e*. The yellow variety may have either black or liver nose leather, depending whether or not gene *b* is present.

Colour	Genotype
Black	A^s–B–E–$ffs^w s^w T$–
Liver	A^s–bbE–$ffs^w s^w T$–
Yellow, black nose	—B–$eeffs^w s^w T$–
Yellow, liver nose	—$bbeeffs^w s^w T$–

Dandie Dinmont Terrier

This is a difficult breed to analyse for coat colour, largely because of the changes which occur between birth and maturity. According to Little (1957), the mustard variety is born as dark sable while the pepper is born as black and tan, both varieties becoming distinctly paler as adults. He concludes that the paling effect is due to the greying gene *G*. He notes that some peppers could be A^s instead of a^t. Following Little, the genotypes of the two varieties may be written: mustard A^y–G–ll and pepper $a^t a^t G$–ll.

Deerhound

Little (1957) considered the colour of these dogs to be brindle, not the usual pattern of black and yellow stripes but a confused mixture of black and yellowish hairs. The longish coarse wiry coat could contribute to the picture. Little states that the stripes can often be seen at birth or if the coat is clipped. The variation in the intensity of the yellow. The genotype is probably A^y–E^{br}–Wh–. It may be noted that the genotype could be a^{sa}–E^{br}–Wh, since the brindling in this case would tend to be more confused than for the former genotype. Plain yellow, sandy and cream varieties are of genotype A^y–E–Wh. A black mask is usually present. It is unknown if the long-haired gene *l* is involved.

Dobermann

The Doberman is the archetypical tan pattern dog, a fact which is highlighted by the recognition of three colours, black, brown or liver and blue, all with the same pattern. The tan parts of the pattern should be rich in colour and sharply defined. The lilac or isabella variety should appear from time to time, since the genes *b* and *d* are present in the breed. These are probably confused with blue, as an off-colour, or are discouraged.

Colour	Genotype
Black	$a^t a^t B\text{–}D\text{–}$
Brown	$a^t a^t bbD\text{–}$
Blue	$a^t a^t B\text{–}dd$
Isabella	$a^t a^t bbdd$

Dunker

This uncommon but interesting Norwegian breed deserves an entry because it is a merle heterozygote. The colour is blue-grey with ragged black patches; ideally, the latter should be well distributed. White spotting of the "collar" grade is present, suggesting that the s^i allele is involved. Genotype: $A^s\text{–}M,ms^i\text{–}$.

Elkhound

The colour of this sturdy dog is very close to that of the wolf. The basic pattern is saddle, but much lighter in the disposition of black pigment, with abundant ticking of light yellow or pale cream thoughout. This is the major factor in lightening the more usual saddle of, say, the Beagle or Foxhound. Little (1957) is willing to accept that the grey Elkhound has the wild type (wolf) gene A which the present author is inclined to equate with the saddle gene a^{sa} (see earlier chapter). A black mask is present. A black variety is known but apparently is rare. Little has described a liver Elkhound ($a^{sa}\text{–}bb$), presumably a rate segregant.

English Cocker Spaniel

A wide variety of colours is permitted in this popular breed; in fact, almost running through the whole gamut of known colours, together with piebald spotting and the presence of known colours, together with piebald spotting and the presence or absence of ticking. The amount of ticking is not extensive and only occasionally becomes a roan. Solid black, black and tan, sometimes called black and red, and chocolate or liver are common. The status of red may be complicated by the presence of both A^y and e. According to Little (1957), the majority of reds are ee, but a proportion may be A^y, because black offspring have been produced from red parents. Apart from errors of mating, the only way this could happen is from mating of $A^y\text{–}E\text{–}$ with $A^s\text{–}ee$, both parents being similar reds. Little noted the occurrence of two blue pups among the results sent in by breeders to the Jackson Laboratory, which is suggestive that the d gene is present as a rarity.

The intensity of colour of the tan areas in black and tans varies from red through tan to cream. This is matched by variation in the red individuals and is probably due to the same cause, namely the presence of minus rufus polygenes. It could also be due to the rare occurrence of a c^{ch} allele, a possibility raised by Little. He also mentions that some white animals have coloured spots only on the ears. These could be an extreme expression of $s^p s^p$ or be indicative of s^w, present either as $s^p s^w$ or $s^{w\,w}$.

The hair is long and is due to the long-haired gene *l*. Many Spaniels show variable amounts of waviness in the coat, usually on the head and legs, but sometimes on the body. Whitney (1947) states that straight hair appears to be dominant to wavy in the Cocker Spaniel, but admits that the variation of waviness is a problem in diagnosing all wavy individuals.

Colour	Genotype
Black	A^s–B–E–llS——
Black, white	A^s–B–E–lls^p–tt
Black, white, ticked	A^s–B–E–lls^p–T–
Black and tan	$a^t a^t B$–E–llS——
Ditto, white (tricolour)	$a^t a^t B$–$Ells^p$–tt
Ditto, white ticked	$a^t a^t B$–$Ells^p$–T–
Liver	A^s–bbE–llS——
Liver, white (tricolour)	A^s–bbE–lls^p–tt
Liver, white ticked	A^s–bbE–lls^p–T–
Liver and tan	$a^t a^t bbE$–llS——
Ditto, white	$a^t a^t bbE$–lls^p–tt
Ditto, white, ticked	$a^t a^t bbE$–lls^p–T–
Red, black nosed	—B–$eellS$——
Ditto, white	—B–$eells^p$–tt
Ditto, white, ticked	—B–$eells^p$–T–
Red, liver nosed	—$bbeellS$——
Ditto, white	—$bbeells^p$–tt
Ditto, white, ticked	—$bbeells^p$–T–

English Setter

This breed occurs in a number of colours. Black, liver and lemon are the basic colours, but these are combined with white spotting and ticking which is so profuse that it may rightly be regarded as roaning. Many individuals show patches of colour typical of piebald spotting, but equally as many others show merely a few small coloured patches on the head and rump, typical of extreme-white. These latter dogs show the greatest effect of the roaned pattern, literally covering the whole body. It is probable that both s^p and s^w occur in the breed. The breed is a long hair and this feature is partly responsible for the roaning, or belton colour as tricolour. Little (1957) has shown that the lemon colour is due to the *e* gene.

Colour	Genotype
Black flecked	A^s–B–E–lls^p–T–
Liver flecked	A^s–bbE–lls^p–T–
Black tricolour	a^ta^tB–E–lls^p–T–
Liver tricolour	a^ta^tbbE–lls^p–T–
Lemon, black nosed	—B–$eells^p$–T–
Lemon, liver nosed	—$bbeells^p$–T–

English Springer Spaniel

Similar to the English Cocker Spaniel (*q.v.*), the present Spaniel is bred in a large number of different colours. These are black and liver, either solid or tan patterned, red or yellow, according to the number of rufus polygenes, and spotted or unspotted with white.

Colour	Genotype
Black	A^s–B–E–llS—
Black, white	A^s–B–E–lls^p–tt
Black, white, ticked	A^s–B–E–lls^p–T–
Black and tan	a^ta^tbbE–llS——
Ditto, white	a^ta^tbbE–lls^p–tt
Ditto, white, ticked	a^ta^tbbE–lls^p–T–
Liver	A^s–bbE–llS——
Liver, white	A^s–bbE–lls^p–tt
Liver, white, ticked	A^s–bbE–lls^p–T–
Liver and tan	a^ta^tbbE–llS——
Ditto, white	a^ta^tbbE–lls^p–tt
Ditto, white, ticked	a^ta^tbbE–lls^p–T–
Red, black nosed	—B–$eellS$——
Ditto, white	—B–$eells^p$–tt
Ditto, white, ticked	—B–$eells^p$–T–
Red, liver nosed	—$bbeellS$——
Ditto, white	—$bbeells^p$–tt
Ditto, white, ticked	—$bbeells^p$–T–

English Toy Terrier

This is a miniature of the Manchester Terrier with the genotype a^ta^t.

Field Spaniel

As in the other Spaniels, there are a large number of recognized colours, except that in the Field Spaniel non-spotted varieties are given preference. Piebald-spotted animals are not disallowed, but nor are they encouraged. The list of genotypes for the breed would read like those for the English Springer Spaniel, to which the reader is referred for details and for the relevant genotypes.

Finnish Spitz

A yellowish-red sable of moderate and symmetrical shading. The coat is long. Genotype: $A^y\text{–}ll$.

Flat-coated Retriever

This breed exists in two colours, black or liver. Their genotypes are $A^s\text{–}B\text{–}$ and $A^s\text{–}bb$, respectively.

Foxhound

This is a typical saddle patterned dog with white spotting of the piebald grade. However, black, yellow or tan coloured animals are making an appearance, again with white markings.

Colour	Genotype
Black	$A^s\text{–}s^p\text{–}$
Yellow	$A^y\text{–}s^p\text{–}$
Saddle	$A^{sa}\text{–}s^p\text{–}$

French Bulldog

The fawns have a black mask and are A^y. The brindles are due to the E^{br} gene and tend to be very dark. In fact, most black animals are brindles with no or very little yellow markings. White spotting is common and varies from white breast marking (Irish type) to almost all-white. It may be assumed that all three spotting alleles are present in the breed.

Colour	Genotype
Fawn	$A^y\text{–}E\text{–}S\text{–}$
Brindle	$A^y\text{–}E^{br}\text{–}S\text{–}$
Fawn, white	$Ay\text{–}E\text{–}s^p\text{–}$
Brindle, white	$A^y\text{–}E^{br}\text{–}s^p\text{–}$

German Shepherd Dog

Several distinctive colour varieties occur in this breed, together with significant variation of the individual phenotypes. The sable may vary from a yellow with little ticking to one which is heavily shaded. The saddle may vary from a light animal with abundant light ticking to one in which the saddle is almost solid black. The black and tan shows the least variation, but even here there is variation, with a diminution of the extent of the tan markings, e.g. about the muzzle by the presence of a black mask. The black mask is shown by most dogs except black, where, of course, it is hidden. The intensity of yellow varies from a golden to pale cream bordering on white for the sable, saddle and tan pattern phenotypes.

The above summary implies that the genes A^y, a^{sa}, a^t and a of the agouti, and c^{ch} of the albino series are present, producing the range of variation. The interesting aspect is the variation of expression shown by a^{sa}. The variation of A^y is not remarkable in that it does not differ from that shown by A^y in other breeds. However, the variation shown by a^{sa} interacting with rufus polygenes deserves closer examination. The saddle refers to the V-shaped pattern on the sides of the dog. It may be a narrow V or a wide V stretching from the shoulders to the flanks. It may be solid black or ticked with variable numbers of yellowish hairs. The V-shaped saddle may be the prime feature, but the pattern also produces characteristic dark marking on the face, across the breast and shoulders, and a dark stripe down the thighs.

The saddle pattern is thus a complex entity and should be visualized as set against a background of yellow pigmentation. The depth of colour varies from a golden to pale cream. The paler the background, the more wolf-grey becomes the saddle pattern. It only requires the appropriate degree of ticking for a simulation of a typical wolf coloration. It is by no means certain that the coloration is simply a simulation, for the a^{sa} allele is probably identical to A. It is only possible to observe in a few breeds, of which the German Shepherd dog is a fine example, how the usual phenotype of A has been modified. Little (1957) has concluded that the wild type gene is probably present in the German Shepherd dog, evidently because of the existence of wolf-like coloured individuals. Unfortunately, Little has not separated a^{sa} from a^t in his analysis of canine coat colours and has to postulate a "wild type" allele (symbol a^w but equivalent to A).

The pale cream or white with dark points which occur from time to time could be A^y animals either combined with a c^{ch} gene (as suggested by Little) or with minus acting rufus polygenes diluting the yellow pigment. The dark points are due to the mask characteristic which would be unaffected by either c^{ch} or the rufus polygenes.

The grey sable, in which the intensity of yellow is reduced, is produced by the chinchilla gene c^{ch}, a gene which degrades yellow but not black pigment. However, the richness of red or yellow can also be diluted by minus rufus polygenes and can produce a grey sable. The two sorts of grey sable may be distinguished by their breeding behaviour. The former will assort distinctly while the latter will produce a graded series of greyish sables. The black and cream or black and silver are produced by the combination of genes a^t and c^{ch} or by the minus rufus polygenes.

Carver (1984) has shown that the all white variety, with dark eyes and skin colouring, is inherited as a recessive to normal coloured. He has concluded that the colour is not produced by either a combination of c^{ch} with a yellow gene or by extreme white spotting s^{w}. Instead, another gene w for recessive white coat colour is proposed.

It is of interest that the solid black of the German Shepherd dog is not due to A^{s}, as seems to be true for most breeds, but to the recessive a allele. This unusual situation was foreshadowed by the early observations of Yentzen (1965) and confirmed by the studies of Willis (1976) and Carver (1984).

A popular breed such as the German Shepherd would be expected to have other genes which are present at low frequencies. For example, Carver has shown that the clear yellow variety (no sable shading) is almost certainly produced by the recessive non-extension gene e. These could be bred from either black and tan or solid black, which would not be possible even with a sable devoid of sable shading. Other rare colour possibilities are chocolate and blue. All of the colours may breed with either short or long coat.

Colour	Genotype
Sable	$A^{y}-C-$
Saddle	$a^{sa}-C-$
Black and tan	$a^{t}-C-$
Black	$aaC-$
Grey Sable	$A^{y}-c^{ch}-$
Black and cream	$a^{t}-a^{ch}-$

German Short-haired Pointer

These dogs are solid liver in colour, with or without extensive white of the piebald grade. The white areas of many animals are liberally sprinkled with liver spots, almost to the point of roaning. On the smooth coat of the short-haired breed, however, the ticking can be seen to be spots, albeit closely set together, rather than a true roan.

Colour	Genotype
Liver	$A^{s}-bbS$——
Liver, white	$A^{s}-bbs^{p}-tt$
Liver, white, ticked	$A^{s}-bbs^{p}-T-$

German Wire-haired Pointer

The discussion for the German short-haired Pointer is fully applicable to this breed as regards coat colour. In fact, in essentials, the only differences between the two breeds is that the wire hair possesses the *Wh* gene for wire hair. Apart from producing a rough resilient coat, the effect of *Wh* is to enhance the effect of roaning. This arises from the nature of the wire-haired coat which is not so smooth or close-lying as that of the short-haired. Thus, there is no real necessity to postulate a gene other than the ticking gene *T*. Despite this, some people have postulated from the general appearance of "roaning" that a roaning, as distinct from *T*, gene must exist, but there is no experimental evidence for this.

Golden Retriever

This breed is a true breeding yellow, but may vary considerably in intensity from a rich golden to pale cream. The colour is probably due to *e*, hence the basic genotype is *eell*. The variation of intensity of yellow is due in part to the effects of rufus polygenes. However, some animals are so pale as to lead to speculation that a chinchilla-like (c^{ch}) gene may be involved. If this is the case, the effect of c^{ch} on *eell* would be to create a tawny-yellow.

Gordon Setter

This is a black and tan pattern dog, with no other colour being permitted. The hair is only moderately long, but it is probably *ll* with modifiers for a close lying coat. The colour of the tan markings is usually red, but Little (1957) writes of the occurrence of animals with a pattern of very light straw. He ascribes this to the action of a c^{ch} gene. If these animals occur as distinct segregants, i.e. no intermediate grades of red or yellow, Little is probably correct, otherwise one must assume that minus rufus polygenes are at work. Both Winge (1950) and Little comment that red animals are not unknown in the breed. If these are bred from ordinary Gordon Setters, their genotype must be $a^t a^t eell$.

Great Dane

The colours present in the breed are fawn (which actually vary from light orange to fawn), brindle, black and blue. The fawns are due to A^y which interacts with E^{br} to produce brindle which, incidentally, should be nicely striped black upon yellow, not too dark, nor a confused brindling. The blues are dilute blacks. Blue brindles occur, but these are barely permitted because of their unattractive appearance. A black mask is a feature of most colours except in dilute fawns or blue brindles where it is bluish, contrasting poorly with the body colour.

The harlequin Great Dane has a resemblance to the merle Collie or dapple Dachshund. In these two breeds, the pattern consists of a mixture of black and bluish areas. However, in the Dane, the mixture is a characteristic pattern of ragged black patches on a white or very light background. An analysis of the heredity of the pattern by Sponenberg (1985) and by O'Sullivan and Robinson (1989) has revealed that it is produced by the merle gene M but in conjunction with another gene H it causes the usual bluish areas to become white. Curiously, the H gene is lethal when homozygous HH. The core genotype of the harlequin pattern, therefore, is $HhMm$ regardless of any other genes which may be present. The pattern of coloured patches of the harlequin appear to be more broken up than those found in the merle. This may be due in part to the greater contrast between the black and white areas but it may be due in part to selective breeding for a well distributed pattern.

Colour	Genotype
Black	$A^s{-}D{-}E{-}hhmm$
Fawn	$A^y{-}D{-}E{-}hhmm$
Brindle	$A^y{-}D{-}E^{br}{-}hhmm$
Blue	$A^s{-}ddE{-}hhmm$
Blue brindle	$A^y{-}ddE^{br}{-}hhmm$
Black harlequin	$A^s{-}D{-}E{-}HhMm$
Blue harlequin	$A^s{-}ddE{-}HhMm$
Merle	$A^s{-}D{-}E{-}hhMm$

Greyhound

The Greyhound is bred in a large number of varieties and all of the common colours are well represented, with the exception of tan pattern and liver. There is little point in discussing any of these, except to comment that the reds may vary through yellow, fawn to cream. This variation is due to the action of the major gene d and modifiers of the rufus group. Little (1957) considers that all three of the white spotting alleles are present in the breed. The variation is considerable, but most of it is accountable to the S^p gene.

Colour	Genotype
Black	$A^s{-}D{-}E{-}S{-}$
Blue	$A^s{-}ddE{-}S{-}$
Black brindle	$A^y{-}D{-}E^{br}{-}S{-}$
Blue brindle	$A^y{-}ddE^{br}{-}S{-}$
Red, yellow	$A^y{-}D{-}E{-}S{-}$
Fawn, cream	$A^y{-}ddE{-}S{-}$
Black, white	$A^s{-}D{-}E{-}s^p{-}$
Blue, white	$A^s{-}ddE{-}s^p{-}$
Black brindle, white	$A^y{-}D{-}E^{br}{-}s^p{-}$
Blue brindle, white	$A^y{-}ddE^{br}{-}s^p{-}$
Red, yellow, white	$A^y{-}D{-}E{-}s^p{-}$
Fawn, cream, white	$A^y{-}ddE{-}s^p{-}$

Griffon Bruxellois

Black, black and tan, and red are the colours found in this breed. A mask is held to be desirable for the red. These little dogs are bred in two coats, smooth and wire hair. Some of the latter give the impression of being long-haired as well as being wiry.

Colour	Genotype
Black	$A^s\text{–}whwh$
Black and tan	a^ta^twhwh
Red	$A^y\text{–}whwh$
Black wire-haired	$A^s\text{–}Wh\text{–}$
Black and tan wire-haired	$a^ta^tWh\text{–}$
Red wire-haired	$A^y\text{–}Wh\text{–}$

Haldenstover

A well-built medium size dog of saddle pattern with piebald grade of white spotting. Genotype: $a^{sa}\text{–}s^p\text{–}$.

Harrier

A smaller edition of the Foxhound (*q.v.*).

Hovawart

Three colour varieties are bred in the breed. These are black, black and gold, and blonde. The black is due to the agouti allele A^s, while the black and tan is produced by the a^t allele. Some, if not many, black and tans have the usual facial markings of the pattern partially obscured by the presence of mask. The blonde is a golden yellow, devoid of sable shading. This fact implies that the colour is due to the non-extension allele *e*, and this has been confirmed by the analysis of breeding records. The coat is long haired (Robinson, 1989*b*).

Colour	Genotype
Black	$A^s\text{–}$
Black and Gold	a^ta^t
Gold	ee

Hungarian Vizslas

A yellow dog of varying shades, from rich sandy yellow to russet gold. Interestingly, the nose leather and lips are brown. Genotype: $A^y\text{–}bb$.

Ibizan Hound

A red or yellow dog with piebald grade of white spotting. Possibly *bb* as judged by nose colour (liver), but this is uncertain. The coat is usually smooth but long- (*ll*) and wire-haired (*Wh*) versions are known.

Colour	Genotype
Red	A^y–L–s^p–$whwh$
Red long hair	A^y–lls^p–$whwh$
Red wire hair	A^y–L–s^p–Wh–

Irish Setter

The redness of this genetically "yellow" dog is outstanding and is the hallmark of the breed. Little (1957) has shown that the absence of black is due to the *e* gene. Although the coat is not overly loose nor full, the abundant feathering gives the impression that the *l* gene is present. If so, the genotype is A^s–*eell*. The rich red colour is due to maximum concentration of plus rufus polygenes.

Winge (1950) notes that black pups have been produced by red parents in the past, while Little (1957) mentions one black among 1197 pups reported by breeders to the Jackson Laboratory. Burns and Fraser (1966) say that information from breeders clearly indicates that very occasionally red Irish Setters do throw black and tan, black and sable offspring. These comments suggest that while the majority of reds are probably A^s–*ee*, a small number may in fact be A^y–*EE*.

Irish Terrier

The Irish Terrier is a yellow with a wire-haired coat. The colour of the yellow pigmentation varies from reddish-yellow to light red and is governed by rufus polygenes. Genotype: A^y *Wh*.

Irish Water Spaniel

This breed is a solid coloured liver of genotype A^s–*bb* but with a long curly coat. Whitney (1957) has published data which suggests that the kinky coat (Whitney's terminology) is inherited as a recessive. Accepting these data as valid, the genotype is A^s–*bbkkll*.

Irish Wolfhound

As in the Deerhound (*q.v.*), Little (1957) considers the brindle colour to be due to the E^{br} gene but with the usual striped pattern broken up.

However, the possibility that the brindle has the genotype a^{sa}–E^{br}– should be considered. These dogs present a dark confused sort of brindling to the eye. The coarse wiry coat would contribute to some extent to the absence of a definite pattern. Blacks, reds, fawns and white are mentioned in the standard. Little suggests that the whites are actually pale cream due to an interaction of the A^y and c^{ch} genes, or to minus rufus polygenes lightening the yellow pigment normally present in A^y animals. It is unknown if the long-haired gene *l* is present.

Colour	Genotype
Black	A^s–E–
Brindle	A^y–E^{br}–
Red, fawn	A^y–E–

Italian Greyhound

The remarks in the section on the Greyhound (*q.v.*) are fully applicable to the present breed, except that brindles or blue-brindles do not occur. This implies that the brindle gene E^{br} is excluded from the breed.

Japanese Chin

The colours here are black and red, both combined with piebald spotting. The reds may be clear of black ticking or display variable amounts of sable shading. The coat is long and of silky texture.

Colour	Genotype
Black and white	A^s–lls^p–
Red and white	A^y–lls^p–

Keeshound

The standard states that these should be wolf-grey or ash-grey. Their phenotype could be formally written as *A*, i.e. as the wild-type genotype. However, the phenotype is saddle (a^{sa}) with minus rufus polygenes causing the normally yellow areas to be pale cream or white. The coat is long. The genotype is *A*–*ll* or a^{sa}–*ll*.

Kerry Blue Terrier

The pups of this are born black, but most change gradually to a blue, which may range from a slate to a light silvery colour. This transformation with age is characteristic of Little's greying gene *G*. A mask may be present in some animals, but not in others. It is possible that the *G* gene cannot dilute the abundant pigmentation necessary to induce a mask to bring about an appreciable change. The coat is long, full and wavy. The genotype may be tentatively written as A^s–*G*–*ll*.

King Charles Spaniel

This breed is essentially a smaller version of the Cavalier King Charles Spaniel (*q.v.*).

Labrador Retriever

Black and yellow are the common colours of this handsome dog, although liver is not unknown. It is almost certain that yellow in the breed is due to *e* (Burns and Fraser, 1966). The variation of intensity of yellow is due to rufus polygenes.

Colour	Genotype
Black	A^s–*B*–*E*–
Liver	A^s–*bbE*–
Yellow, black nosed	——*B*–*ee*
Yellow, liver nosed	——*bbee*

Lakeland Terrier

This breed is akin to the Airedale Terrier but occurs in a wider range of colours. The listed acceptable colours are black, blue, liver, black and blue saddle, red, wheaten and red grizzled. The red grizzled could be a light sable, but is probably a light saddle. The coat is a wire hair.

Colour	Genotype
Black	A^s–*D*–*Wh*–
Saddle	a^{sa}–*D*–*Wh*–
Blue	A^s–*ddWh*–
Blue Saddle	a^{sa}–*ddWh*–
Red	A^y–*D*–*Wh*–

Large Munsterlaender

This medium-sized dog is white with a few patches of black and a variable amount of ticking. The ticking may range from a few spots to abundant ticking so that the white areas have a strong resemblance to roan. The coat is long. Genotype: A^s–lls^p–T–.

Maltese

Pure white is the preferred colour in this breed, although slight yellow markings are permissible. Little (1957) suggests that the whiteness is due to the extreme-white allele s^w rather than a combination of a yellow gene (A^y or e) with chinchilla (c^{ch}) and/or minus rufus polygenes. He states that the yellow patches occur about the head or body, which would be consistent with the s^ps^p suggestion. However, if the yellowness occurs as a more general suffusion, say along the spine, the suggestion that the whiteness is due to A^y–c^{ch}– or ch–ee is more probable.

Manchester Terrier

This is one of the archetypal black and tan breeds. Genotype: a^ta^t.

Newfoundland

The two recognized colours are black and liver or chocolate. Whitney (1947) mentioned that red and blue Newfoundlands are known, but presumably these occurred as rare segregants. The Landseer is a Newfoundland with piebald spotting. Whitney (1947) mentioned that a Newfoundland dog transmitted ticking (T) to the white areas of his Landseer puppies. The coat is long.

Colour	Genotype
Black	A^s–B–D–S–
Liver	A^s–bbD–S–
Blue	A^s–B–ddS–
Landseer	A^s–B–D–s^p–

Norfolk Terrier

The usual colour is a A^y yellow of a light reddish shade, although lighter colours are known. Black ticking may occur on the head, tail and sides of the body, whence the animal is said to be grizzled. Black and tan individuals also occur in the breed. The coat is seemingly wire hair.

Colour	Genotype
Red	A^y–Wh–
Grizzled	A^y–Wh–
Black and tan	$a^ta^t Wh$–

Norwegian Buhund

Colours may be wolf-grey, light red to wheaten yellow and black. Masks may be seen in some animals.

Colour	Genotype
Black	A^s-
Yellow	A^y-
Wolf-grey	$A-$

Norwich Terrier

This little terrier appears to be identical to the Norfolk Terrier (*q.v.*) in respect to coat colour and type.

Old English Sheepdog

This is another breed which Little (1957) considers to have the greying gene G. The characteristic colour would bear out the suggestion. Little admits, however, that some individuals could be ordinary blues as produced by the dilute factor d. It is difficult to discover if the merle gene M is present or not, although Little considers the possibility. The majority of animals are white marked of the "collar" grade or lighter although the standard says that non-spotted are permitted. The coat is long, full and shaggy, an aspect which is a complication in analysing the colour. The genotype of the typical individual may be provisionally written as $A^s-G-lls^v-$.

Papillon

This is one of the most dainty of all dogs. They are white with patches or spots of black, red or tan pattern, the latter being the tricolour variety. The coat is long, silky and thrown up as frills.

Colour	Genotype
Black	A^s-lls-
Red	A^y-lls-
Black and tan	$a^t a^t lls^p-$

Pekinese

Pekinese are bred in a number of colour varieties, among which the red and sable are the most popular. Both of these colours are produced by the dominant yellow gene A^y, with the addition of sable shading for the latter.

A black mask is also present. The sable may occur with either red or fawn ground colour; the latter being the grey sable of genotype $A^s c^{ch}$. The fawn or cream appears to have the same genotype as the grey sable but are devoid of the sable shading. According to the distribution of the white pattern, this is produced by the piebald gene s^p. It is of interest that albino individuals—white, with red or pink eyes—have turned up in the past. The coat is exceptionally long and flowing.

Colour	Genotype
Red	A^y–C–lls–
Cream	A^y–c^{ch}–lls–
Sable, red	A^y–C–lls–
Sable, grey	A^y–c^{ch}–lls–
Red, white	A^y–C–lls^p–
Sable, red, white	A^y–C–lls^p–
Sable, grey, white	A^y–c^{ch}–lls^p–

Pinscher

Several well-known colours are bred in this breed: red, black, black and tan, chocolate, blue and blue and tan.

Colour	Genotype
Black	A^s–B–D–
Chocolate	A^s–bbD–
Blue	A^s–B–dd
Black and tan	$a^t a^t B$–D–
Chocolate and tan	$a^t a^t bbD$–
Blue and tan	$a^t a^t B$–dd
Red (black nose)	A^y–B–dd
Red (liver nose)	A^y–bbD–
Cream (slate nose)	A^y–B–dd
Cream (pale nose)	A^y–$bbdd$

Pointer

The usual pointer is a white dog with coloured patches occurring on any part of the body. The patches may be coloured black, liver or yellow. It is usual to assume that yellow in the pointer is due to the *e* gene (e.g. Little, 1957). If this is so, then the genotypes may be written as:

Colour	Genotype
Black, white	A^s–B–E–$s^p s^p$
Liver, white	A^s–bbE–$s^p s^p$
Yellow, white, black nose	A^s–B–$ees^p s^p$
Yellow, white, liver nose	A^s–$bbees^p s^p$

However, a few yellows show a vestige of a mask and sooty ears, features which are suggestive of A^y yellow than e. If so, then A^y– and E– should be substituted for A^s and ee in the yellow genotypes.

The nature of the white markings is typical of that expected for the s^p allele and this is shown in the above genotypes. However, Little (1957) has observed that Pointers in the United States are whiter than those in Britain or Europe. While agreeing that British and European animals are $s^p s^p$, he suggests that many in the United States are $s^w s^w$.

In some dogs the white areas are clear, but the majority are covered in coloured ticking, the colour corresponding to that shown by the patches. The ticking can vary from a few flecks here and there to extensive flecking. This is due to the ticking gene T, and this symbol must be added to the above genotypes to represent the ticked varieties.

Pomeranian

Many colours exist in the breed as may be produced by the primary colour genes. Black, liver and blue, both in varying shades, red, sable in varying shades, and cream. On occasion, piebald spotting (s^p) may be seen, combined with any of the above colours. The coat is exceptionally long, full and silky in texture.

Colour	Genotype
Black	A^s–B–D–ll
Liver	A^s–bbD–ll
Blue	A^s–$Bddll$
Red (black nose)	A^y–B–D–ll
Red (liver nose)	A^y–bbD–ll
Sable (black nose)	A^y–B–D–ll
Sable (liver nose)	A^y–bbD–ll
Cream (slate nose)	A^y–B–$ddll$
Cream (pale nose)	A^y–$bbddll$

Poodle

The poodle at first sight would seem to present a bewildering range of colours. Most of the standard colours are present, black, liver, blue and light red. In addition, there is the greying gene G which modifies all of the above colours to a lesser or greater degree. That is, the colour will depend upon the amount of greying which can result from the presence of G. Remember that animals with G are those which are born normally coloured but change with maturity. Little (1957) believes that the reds and creams are e (rather than A^y), but no evidence seems to be available on this point. The coat is long and wiry and is contributory to the profusion (as well as confusion) of the various shades.

Colour	Genotype
Black	A^s–B–D–E–gg
Blue	A^s–B–ddE–gg
Liver	A^s–bbD–E–gg
Lilac-grey	A^s–$bbddE$–gg
Bluish-grey	A^s–B–D–E–G–
Light blue	A^s–B–ddE–G–
Brownish-grey	A^s–bbD–E–G
Brownish-silver	A^s–$bbddE$–G–
Light red (black nose)	—B–D–$eegg$
Light red (liver nose)	—bbD–$eegg$
Cream (slate nose)	—B–$ddeegg$
Cream (light nose)	—$bbddeegg$
Apricot to cream	——D–eeG–
Cream to off-white	——$ddeeG$–

Pug

The black and fawn are ordinary A^s and A^y individuals, complete with a pronounced black mask for the latter. The silver fawn also possesses a black mask, but the body hairs are a light cream. Little (1957) credits the diluting of the yellow pigment to a chinchilla gene c^{ch}, which such a gene would certainly do. However, the dilution could be due to minus acting rufus polygenes.

Colour	Genotype
Black	A^s–C–
Fawn	A^y–C–
Silver Fawn	A^y–c^{ch}–

Pyrenean Mountain Dog

The breed is an instance of extreme-white spotting and is demonstrated by the occurrence, and acceptance by the standard, of almost all-white dogs but for coloured patches on the head, ears and base of the tail, exactly those positions where the coloured areas persist the longest in the progression to the all-white individual. According to the data of Little (1957), only about 24 per cent of Pyrenees are completely devoid of coloured areas.

The colour of the patches will depend upon the genes present in the animal, masked by the white coat. Those with wolf-grey markings of various shades are probably a^{sa}–, with variable numbers of paler rufus polygenes, while those with lemon patches are probably A^y–, again with minus rufus polygenes lightening the colour. The coat is full and long, and presumably genetically ll. The genotypes of the main varieties may be represented as a^{sa}–lls^ws^w and A^y–lls^ws^w. In appearance, these will be black-eyed whites due to the overriding effect of s^ws^w.

Rhodesian Ridgeback

This interesting dog is characterized by a reversal of direction of hair growth along the spine, a peculiarity which gives the breed its name. The ridge must be clearly defined, symmetrical and tapering towards the tail. The inheritance of the hair reversal has been shown to be due to a recessive gene *ds*. The colour is a yellowish-red usually accompanied by a black mask. Genotype A^{y}*–dsds*.

Rottweiler

The Rottweiler is one of the archetypal black and tan breeds of genotype $a^{t}a^{t}$. The intensity of the tan markings vary but not to the same extent as in many other breeds.

Rough Collie

The Rough Collie is bred in a limited number of colours. There is A^{y} yellow in the form of sable of varying shades or with no shadings (whence they are called clear sable) and black and tan pattern. The merle variety is a patchwork of dark and light colours and is produced by the heterozygote *Mm*. The merle pattern is not particularly distinctive in the clear sable or even lightly shaded sable, but emerges as a patch-like mottling of black and blue-grey in the dark sable and black and tan. A typical "collar" grade of white occurs in the Collie and is probably due to the s^{i} gene, rather than s^{p}, because the white markings are too regular to be due to s^{p} (Little, 1957). Little mentions that dilute (*d*) individuals are known, occurring as pale yellow, bluish sables and blue and tans. The coat is long and flowing, especially on the shoulders and chest. The powder puff (*pp*) and CN dilution (*cn*) genes have been reported for the Collie, but it seems that these are not in the main stream of the breed.

Colour	Genotype
Sable	$A^{y}-D-llmms^{i}-$
Black and Tan	$a^{t}a^{t}D-llmms^{i}-$
Sable and merle	$A^{y}-D-llMms^{i}-$
Black and tan merle	$a^{t}a^{t}D-llMms^{i}-$
Blue sable	$A^{y}-ddllmms^{i}-$
Blue and tan	$a^{t}a^{t}ddllmms^{i}-$
Blue sable merle	$A^{y}-ddllMms^{i}-$
Blue and tan merle	$a^{t}a^{t}ddllMms^{i}-$

Saluki

A great range of colours occur in this breed, black, black and tan, sable, yellow and pale cream, together with piebald spotting and ticking. Little (1957) suggests that even gene *e* occurs in the breed. The coat is smooth and silky and occurs in two versions, without or with an abundant fringe. The former is probably short-haired *L*– while the latter is long-haired *ll* but with minimum expression on the body. The pale cream varieties would be that expected for a combination of A^y and c^{ch} or to the action of minus rufus polygenes upon the A^y– phenotype.

A representative selection of colours and their genotypes may be shown as follows:

Colour	Genotype
Black	A^s–S–tt
Black and tan	$a^t a^t S$–tt
Sable	A^y–S–tt
Yellow	A^y–S–tt
Black, white	A^s–s^p–tt
Black and tan, white	$a^t a^t s^p$–tt
Sable and white	A^y–s^p–tt
Black, white, ticked	A^s–s^p–T–

Samoyed

The Samoyed is a black-eyed white, the absence of pigment being due to homozygosity of the s^w allele. According to Little (1957), approximately 9 per cent of Samoyeds show small spots of pigmentation of not more than 10 per cent of the body surface. The presence of these coloured areas is excellent evidence that the Samoyed of the United States at least, where Little's data were collected, is $s^w s^w$.

Tjebbes and Wriedt (1927) have reported results suggestive that the Samoyed has the genotype A^s–$ees^w s^w$. These authors remarked that some individuals show small areas of yellowish pigment on the ears, on the back near the tail and on the extremities. Crosses of a Samoyed with a wolf-grey Alsatian, heterozygous for a^t, gave five black young, all of which possessed white markings (grade not stated). Another cross with a brindle heterozygous for *e* gave one brindle and two pups which are described as pale chamois. These latter are interpreted as pale coloured yellows. The black parts of the brindle pattern was normal in colour but the yellow was very pale. This implies that the Samoyed carries minus rufus modifiers capable of appreciably lightening the yellow pigment.

Tjebbes and Wriedt attribute the white fur, and particularly the pale yellow colouring, to an albino allele, but this is unlikely. The only possibility that could involve an albino allele, in the form of chinchilla, is the cream or biscuit variety. The genotype for this colour could be $c^{ch}c^{ch}ee$ or *ee* with rufus modifiers severely lightening the colour, as discussed in the section on the Golden Retriever (*q.v.*).

The coat is full and long, and presumably *ll*. The provisional genotype of the Samoyed may be given as A^s–$eells^ws^w$.

Schnauzer

This breed appears to occur in two colours, solid black and another described as "pepper and salt". The occurrence of the latter is intriguing. There is variation and some specimens defy ready classification, especially the very dark individuals with minimal light ticking. However, many give the impression of saddle animals but with the saddle spread all over the sides of the individual. It is known that this is one of the manifestations of saddle. The yellow pigment is very pale, due presumably to minus rufus polygenes. The Schnauzer escapes being called wolf-grey because of the general dark appearance. The coat is long and probably wiry. This is a factor in disguising the true nature of the colour. The genotype of the black is seemingly A^s and that of the "pepper and salt" is provisionally a^{sa}.

Scottish Terrier

The brindle colour of this breed is probably similar to that of the Cairn Terrier. That is, it is due to the E^{br} gene, presenting to the eye a mixture of black and yellow (confused brindling as opposite to striping), a result of the long wiry coat. The depth of colour of the yellow hairs vary from yellow to pale cream under the influence of minus rufus polygenes. The wheaten of A^y yellow show a similar variation of intensity due to the same polygenes. The brindle varies from dark to light independently of the background colour, according to the proportions of black and yellow hairs. The darkest of these could easily pass as black: hence it is uncertain if the A^s gene is present or not.

Colour	Genotype
Brindle	A^y–E^{br}–$llWh$–
Wheaten	A^y–E–$llWh$–

Sealyham Terrier

The white coat of this little terrier is due to the extreme-white allele s^w, as shown by the common occurrence of coloured patches on the head and ears. The data of Little (1957) indicate that as many as 60 per cent of dogs may be spotted. Little states that the spots are too small to enable reliable indentification of the underlying colour genes. Nevertheless, it is probable that the genotypes of the Sealyham are $A^y\text{–}lls^w s^w$ or $a^{sa}\text{–}lls^w s^w$.

Shetland Sheepdog

The colour and coat genetics of this graceful sheepdog are identical to those for the Rough Collie (*q.v.*).

Skye Terrier

Little (1957) briefly discusses this breed but arrives at no definite conclusion, except to note that the colour genetics are probably similar to those of the Cairn Terrier (*q.v.*). He states that some individuals have a light topknot, similar to that shown by the Bedlington Terrier. It seems possible, if not probable, that the grey varieties owe their colour to the greying gene *G*.

Small Munsterlaender

This is a Spaniel-type dog, extensively white with chocolate or liver patches. The white areas are profusely ticked, almost to the stage of appearing as roan. The coat is long. Genotype: $A^s\text{–}bblls^p\text{–}T\text{–}$.

Smooth Collie

The coat colour genetics of this form of Collie are identical to those for the Rough Collie (*q.v.*).

Smooth Fox Terrier

The most common colours seem to be black and saddle. In the latter the saddle is V-shaped and solidly black, while the yellow areas are light red or golden. The golden and white animals could be either A^y or *e*; or a^{sa}, where the colour spots happen to occur only in yellow areas. The white spotting is light piebald.

Colour	Genotype
Black	$A^s\text{–}s^p\text{–}$
Saddle	$a^{sa}\text{–}s^p\text{–}$

Soft-coated Wheaten Terrier

The coat is long, full and of a soft texture. The colour is wheaten, which is likened to ripening wheat. The light yellow colour is due in part to minus rufus polygenes and in part to the long hair which exposes the lighter coloured undercolour. The juvenile coat may show dark ticking, but these usually disappear with age and are indicative of the A^y gene. Genotype: A^y–*ll*.

St. Bernard

The St. Bernard appears to be a sable with variable amount of sable shading. A pronounced black mask is invariably present. A white pattern of the "collar" grade is characteristic of the breed and is probably a manifestation of homozygous s^is^i. However, individuals showing piebald grade spotting typical of s^p are not uncommon, hence this allele could also be present. The coat may be either short (smooth) or long (rough). The genotypes of the two forms are A^y–*L*–s^i and A^y–*ll*s^i–, respectively.

Staffordshire Bull Terrier

The more well-known colours are black, red and brindle. These are due to well-established genes. Blue, fawn and blue-brindled are not unknown and arise from the action of the *d* gene upon the former three colours.

Colour	Genotype
Black	a^s–*D*–*E*–
Red	A^y–*D*–*E*–
Brindle	A^y–*D*–E^{br}–
Blue	A^s–*ddE*–
Fawn	A^y–*ddE*–
Blue-brindle	A^y–*dd*E^{br}–

Sussex Spaniel

This attractive breed is basically a solid chocolate or liver long-coated dog of genotype A^s–*bbll*.

Weimaraner

This breed is uniquely coloured, being described in the standard as silver grey. The colour arises from the unusual combination in dogs of the *b* and *d* genes. Both short- and long-haired versions are bred, although the latter are uncommon. The genotypes are A^s–*bbddL*– and A^s–*bbddll*, respectively. The A^s gene is probably invariably homozygous in the breed.

Welsh Corgi, Cardigan

This delightful little animal occurs in several colours, sable, brindle, black and tan and blue merle are among the more common. The majority of animals show white spotting, restricted to the stomach, breast and foreface, which is typical of Irish spotting (s^i). Little (1957) remarks that individuals with piebald amount of white have been observed, indicating that s^p may also be present. Merle animals show a patchwork of black and bluish markings.

Colour	Genotype
Sable	$A^y{-}E{-}mm$
Brindle	$A^y{-}E^{br}{-}mm$
Black and tan	$a^ta^tE{-}mm$
Merle	$a^ta^tE{-}Mm$

Welsh Corgi, Pembroke

This breed is similar to the foregoing, but probably not bred in so many colours. Sable of various shades and black and tan are those usually found. The amount of white is restricted to the stomach, breast, chest, forelegs and facial stripe or spot.

Colour	Genotype
Sable	$A^y{-}s^i{-}$
Black and tan	$a^ta^ts^i{-}$

Welsh Springer Spaniel

The nose leather of this red and white long-coated breed is liver or flesh, indicating that it is *bb* in constitution. The white pattern is representative of piebald spotting. Little believes that the red colour is due to *e*, hence the genotype may be given as $bbeells^p{-}$.

Welsh Terrier

The Welsh Terrier seems to be a smaller edition of the Airedale Terrier as regards coated colour and type. It differs slightly in that the saddle is not so solidly black. In fact, some light ticking is permissible, when the colour is called grizzled. Genotype: $a^{sa}{-}Wh$.

West Highland White Terrier

Little (1957) has suggested that the white coat of this breed is due to the interaction of A^y and c^{ch}. The genotype $A^y\text{–}c^{ch}c^{ch}$ would be expected to produce a pale cream coloration or even white if there has been selection for pure white as stipulated by the standard. This is a surprising suggestion, since it might be imagined that the breed would have the same genotype as the Sealyham Terrier (*q.v.*), but evidently Little is influenced by his comment that occasional animals have pale cream tinges in whole individual hairs or at the base of white-tipped hairs. This is the sort of evidence to establish the existence of the $A^y\text{–}c^{ch}c^{ch}$ genotype, admittedly, but it may be wise to reserve judgement for the moment. However, if the white appearance is due to s^ws^w, selection has carried the "whitening" further in the breed, since head-spotted specimens appear to be extremely rare or non-existent. This is possible if the selection is intense enough. The coat is coarse but long and presumably of genotype *ll*.

Whippet

Whippets are but miniature Greyhounds and the genetics of coat colour are the same in each. The reader is referred to the remarks in the section on Greyhounds.

Wire Fox Terrier

The coat colours of this breed are identical to those of the Smooth Fox Terrier, to which the reader is referred. The coat is of the wiry type.

Colour	Genotype
Black	$A^s\text{–}s^p\text{–}Wh\text{–}$
Saddle	$a^{sa}\text{–}s^p\text{–}Wh\text{–}$

Woolly-haired Griffon

This breed deserves to be mentioned because of its unusual colouring. This is described as leaf brown or that of a dead leaf. The animal appears to be a light coloured solid liver with a full long coat. The long coat assists in the creation of the desired colour by exposing the lighter brown under-colour of the hairs. Its genotype is probably $A^s\text{–}bbll$.

Yorkshire Terrier

This is a saddle pattern breed of rather unusual colouring. The saddle covers most of the sides of the body, but although initially black, pales to a steel-blue. Little (1957) suggests that the colour may be due to the greying gene *G*. However, this suggestion is speculative and it is possible that the colour is partly the result of selection for long straight hair with the maximum exposure of a slate blue undercolour.

8.

Abnormalities

THAT ABNORMALITIES turn up from time to time is one of the hazards of animal breeding. Most often these result from accidents of embryonic development over which the breeder has no control. Their occurrence may be distressful, but they have to be faced up to. A few genetic anomalies may well occur side by side with the chance defects, but so long as the former occur as isolated events, it is impossible to distinguish one from the other.

One of the first clues that an anomaly may be genetic, rather than a chance event, is that the anomaly recurs in different litters of a group of related animals. The incidence is "familial" as the phenomenon is usually termed. The term familial does not necessarily mean genetic. A group of related dogs are often fed and housed identically. That is, they are exposed to the same environment and there may be a factor in the environment causing the anomaly. Conceivably, this could be an unsuspected dietary deficiency or a toxic agent. In this respect, the term familial is ambiguous; it merely means that the anomaly "runs in families and may be genetic, but the situation warrants closer investigation". The next step is to establish the mode of inheritance. When this is accomplished, the anomaly can be said to be truly genetic.

It is important to view the incidence of genetic anomalies in perspective. Every breed is not riddled with anomalies, just waiting to surface. On the contrary, most anomalies occur only rarely, being a problem of the individual stud or blood line. Within recent years only a few anomalies have perhaps attained high levels of incidence: e.g. hip dysplasia, progressive retina atrophy and Collie eye anomaly. In each instance there were special reasons for the high incidence, the primary one being that the anomaly is not immediately obvious. Circumstances can indeed conspire to increase the incidence of a given anomaly. This aspect will be discussed anon. Whenever this heppens, there will be adverse comment from the public and the reputation of the breed will suffer. It behoves all breeders who care for the breed to take remedial action.

One of the useful preliminary enquiries into the possibility that a given anomaly may be genetic is that of breed differences of incidence. If the anomaly occurs at significantly different frequencies between breeds, this fact may be taken as *prima facie* evidence of a genetic influence. Analyses of this nature are invaluable in indicating which breeds are currently experiencing a high incidence ("are at risk") of an anomaly. Measures may be taken to minimize the risk and to eliminate the anomaly. If the defect cannot be eliminated, it should be possible to reduce it to a low level. The following breed surveys are worthy of attention on account of their comprehensiveness: hip dysplasia (Priester and Mulvihill, 1972), ocular defects (Priester, 1972b), patella luxation (Priester, 1972a) retinal atrophy (Priester, 1974), mammary tumours (Priester, 1979) and umbilical and inguinal hernias (Hayes, 1974).

No attempt will be made to give lists of breeds in which certain anomalies have appeared. In the few cases where breeds are mentioned, the intention is to aid identification of the anomaly, not to castigate the breed in any way. For those interested, extensive lists have been published by Foley *et al.* (1979) and Hutt (1979). The only useful function of such lists is to provide guidance if a given anomaly has appeared previously. Except for special cases, the lists are a doubtful guide to anomalies which might be expected to occur in the breed. The great majority of anomalies are rare and could occur in any breed. It is probable that they are confined to one blood line, rather than the breed as a whole. Once sighted and identified, they can be eliminated. Therefore, it seems unnecessary to cast a blight over the breed as a whole. This is not to suggest that no account should be taken of records of genetic anomalies. On the contrary, such records are essential for the rapid identification and comparison of future discoveries. The point is that a given anomaly may occur in any breed at any time.

Categories of Lethal Genes

The genetic anomaly may appear in one of the number of guises. Many genetic texts discuss what are called "lethals"; genes which cause death of the individual. A distinctive type of lethal is that which causes death prior to birth. The dog does not have a well-known case of such a lethal, but that other domestic carnivore, the cat, does. This is the Manx or tailless breed. The Manx cat is produced by a dominant gene *M* and all Manx cats are heterozygous *Mm*, since the homozygous *MM* dies in the uterus. This means that in breeding Manx to Manx, a 2:1 ratio is obtained instead of the usual 3:1 (see Fig. 30). Thus, the Manx cat itself can never be obtained as a true-breeding animal. These two items invariably indicate the presence of a lethal gene; a "prenatal lethal" to describe it precisely.

There is a common type of lethal gene which produces a deformed pup at birth, the "congenital lethal". One has to be careful not to jump to the

Germ-cells from Manx

Germ-cells from Manx	*M*	*m*
M	*MM* Dies	*Mm* Manx
m	*Mm* Manx	*mm* Normal

FIG. 30. The progeny from the mating of two Manx cats. The homozygous *MM* embryo dies in the uterus and the expected ratio of Manx and normal tailed cats is 2:1.

conclusion that all congenital anomalies are genetic. This is not so, because many accidents of development result in congential deformities. On the other hand, many cases are due to lethal genes. A good example is the bird tongue gene, *bt*, the effects of which are illustrated by Hutt (1979). Homozygous *btbt* pups lack the ability to suckle and die within a few days. Although the *bt* gene apparently produces only a minor physical defect, it causes the death of the pup and is a congenital lethal.

Many lethal genes are known which cause not an immediate death but one which overtakes the individual after an interval of weeks, months or even years. Many of the genes attacking the nervous stystem are of this type. For some weeks or months the young pup displays normal behaviour, but this is followed by ataxic symptoms which, as often as not, steadily worsen until death ensues. Other examples of this "delayed action" type of lethality will be found in the following pages of this chapter. The haemophiliac genes can be both congenital or delayed. If the haemophiliac pup can survive being born, it may live for weeks or months before an accident causes it to bleed to death.

Not all lethal genes need cause the death of their recipients. Many do so, but others may only cause death in a proportion of cases. These are sometimes termed "semi-lethals". The distinction may seem a matter of convenience, but it does highlight the wide spectrum of effects induced by such genes.

There is a special category of genes which do not actually cause death *per se*, but they can impair the normal functioning of the individual so that it becomes more than usually susceptible to other threats of life. The *cn* gene which engenders cyclic neutropenia may be cited as an example. Dogs of *cncn* genotype grow more slowly and eventually die from infectious disease. Therefore, while *cn* does not cause death, it is indirectly responsible.

Such genes could be called semi-lethal or "subvital genes". The minor genes causing inbreeding depression may be said to be "subvital polygenes".

Incidence of Anomalies

All genetic anomalies arise from gene mutation, not immediately but at some time in the ancestral past. The majority of anomalies are due to recessive genes. This explains in large measure how the anomaly may remain hidden. If the causative gene is rare (and most are, it may be reiterated), the overwhelming proportion of matings will be between homozygous normal and heterozygous individuals. In breeder's terminology, between normal and "carrier" animal, carrier meaning a dog "carrying" a recessive gene for an anomaly. These matings will not produce an anomalous pup; thus the gene will remain hidden and passed on from one generation to the next in the heterozygotes.

If one is wondering how this can be, consider the genes (two to each individual) being freely passed on in a stud of, say, ten dogs. The total number of genes is $10 \times 2 = 20$. This totality of genes is known as the "gene pool". Thus, if there are n dogs in a breeding group or breed in any one generation, the total number of genes, or the gene pool, will be $2n$. Now suppose that a proportion x are a gene for an anomaly. Assuming no inbreeding, the chances of two "anomalous" genes meeting up is $x \times x = x^2$. Thus, if $x = 1$ per cent, the chances of an anomalous pup is 0.01 per cent, one in ten thousand in other words. These are long odds, one must admit, and any anomaly occurring at this frequency would scarcely be considered to be genetic in origin.

Now consider what is happening to the heterozygous or carrier animals in the gene pool. $1 - x$ will be the proportion of normal genes. Thus, the chances of normal and anomalous genes meeting up will be $(1 - x) \times x$ plus $x \times (1 - x$ which equals $2x(1 - x)$. This is the proportion of carrier individuals. Therefore, when $x = 1$ per cent, the proportion will equal 1.98 per cent. In other words, the occurrence of carriers will be 1.98 times as frequent as the number of anomalous individuals.

If the carriers are indistinguishable from normals, their presence will be undetectable and, what is more significant, unsuspected. The point of these calculations is to emphasize that while the incidence of a recessive anomaly may be low, the incidence of carriers may be much higher. The same fact is shown by the two curves of Fig. 32. Although the percentage of the anomaly soon falls to a low level, the fall in the percentage of carriers is slower. This is a fundamental reason why some anomalies seem to be so persistent. The relevant gene persists in the population in carrier individuals, only surfacing when two of these are mated together fortuitously.

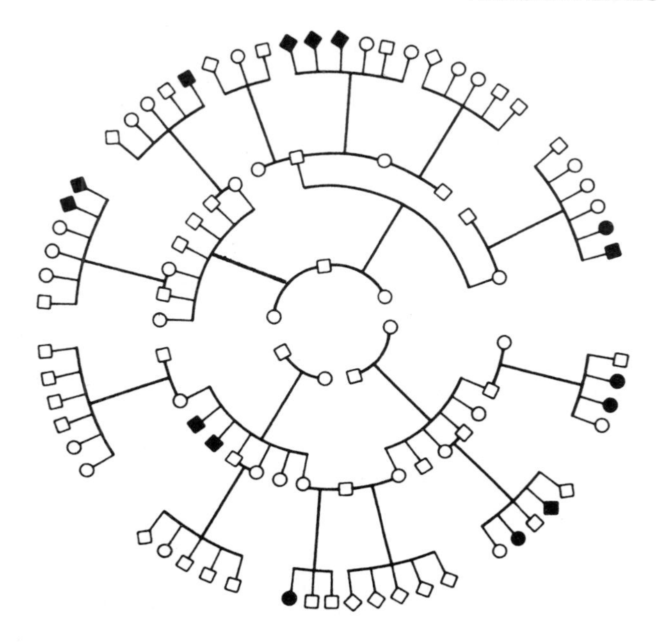

FIG. 31. A typical pedigree chart showing the occurrence of a recessive anomaly (black symbols) in different litters and the genealogical relationships between parents and litters.

Detection of Anomalies

The first sign that an anomaly could be genetic is that several occur in litters of related parents; or, as some people like to express it, the affected pups can be traced back to common ancestors. Figure 31 illustrates a common situation. In this genealogical chart, open squares and circles represent normal males and females, respectively, solid squares and circles represent affected males and females, respectively, and open and solid diamonds represent normal and affected pups of unknown sex. This symbolism is standard and should always be used.

Since the anomaly is bred only from normal parents, it is clearly behaving as a recessive. The only means of discovering heterozygotes is when the dog or bitch produces an anomalous pup. Namely, all litters will have one or more affected pups. Summing the numbers of normal and affected pups gives the ratio of normal:affected as 27:15 or 1.8:1. This is a far cry from the expected 3:1. How could this be? The answer is that litters consisting entirely of normal pups, although bred from carriers, would not be included and the total number of normal pups would be deficient to this extent.

TABLE 13

Expected Number of Pups of the Recessive Phenotype for Small Litters

All litters are observed, but only those with at least one recessive are analysed.

Litter size	Number of expected pups (A) 1:1 ratio	(B) 3:1 ratio
1	1	1
2	1.33	1.14
3	1.71	1.30
4	2.13	1.46
5	2.58	1.64
6	3.05	1.82
7	3.53	2.02
8	4.01	2.22
9	4.51	2.43
10	5.00	2.65
11	5.50	2.87
12	6.00	3.10
13	6.50	3.33
14	7.00	3.56
15	7.50	3.80
16	8.00	4.04

However, given the number of pups for each litter within which at least one affected has been observed, it is possible to allow for the non-inclusion of all normal pups. The expected number of affected pups will vary with the size of litter and the appropriate expectations are given in Table 13. Column A is for matings of normal with affected and column B is for matings of normal with normal. B is the relevant column in the present instance. There are eight litters in which at least one affected has appeared. The procedure is to write down the number of affected animals expected for each litter, according to the litter size as shown by column B. Thus, for a litter of 6, 1.8 affected pups will be expected, for a litter of 4, 1.5 pups, and so on. These expectations are then summed to yield a total of 13.4 for the 8 litters (Table 14). Thus, the observed number of 15 affected is almost exactly the number expected.

It is by the use of procedures such as the above that the mode of inheritance of an anomaly can be analysed. The example is illustrative of simple recessive inheritance. In the case of an anomaly which is inherited as a dominant, the distinguishing feature would be that one parent, at least, always shows the anomaly, unless, of course, the anomaly is not invariably manifested, in which event the pattern of inheritance would appear curious. However, it is still possible to analyse such pedigree charts, but the conclusions are apt to be less precise.

TABLE 14

An Example of the Arithmetic for Working Out the Expected Number of Recessive Pups Shown by the Pedigree Chart of Fig. 31

Litter size	Expected number of recessives	Numbers in example	
		Normal	Recessives
3	1.3	2	1
4	1.5	2	2
5	1.6	4	1
5	1.6	3	2
6	1.8	3	3
6	1.8	4	2
6	1.8	4	2
7	2.0	5	2
	13.4	27	15

Occurrence of Anomalies

It may be wondered how an anomaly comes into being and how it may spread throughout a breed? There are two routes by which an anomaly, may arise. The first is by gene mutation, the second by breeders breeding to an ideal which encourages the animal to develop an anomaly.

Mutations may occur at any time or place, completely at random, in mongrel or pedigree dogs with fine impartially. Mutants may be harmless, such as those giving the basic coat colours, or anomalous, such as the many genetic defects which are coming to the notice of breeders. The reason why so many mutations are harmful is simply that they disrupt a smoothly functioning physiological entity, namely the body of the dog. Somewhere in the embryonic pup a biochemical process fails to function as efficiently as it should and a chain of events is set in motion which culminates in an anomaly. Fundamentally, mutation is as simple as that, solely because of a change in the gene controlling the process. What causes the gene to change? For most practical purposes, the answer is chance; a mistake in the self-copying mechanism of gene reproduction.

A mutation to a dominant gene can produce an anomaly which is usually quickly eliminated. There are exceptions, however. For example, the merle gene *M* produces a white dog with ocular anomalies and deafness when homozygous *MM*, but a merle or harlequin colour when heterozygous *Mm*. Some people find the merle attractive and breed purposely for it. With the application of a little genetic knowledge it is possible to breed merle dogs without the appearance of the undesirable white form with eye troubles. One should breed merle only with normal (non-merle) partners. In this manner the *M* gene persists in the dog population.

Another situation by which a dominant gene can persist in the population is when the clinical effects are not noticeable until well into breeding before anyone is aware. Once the alert has been sounded, however, one can observe closely the descendants for any signs of the anomaly. An attempt should be made by veterinarians or clinicians to detect the onset of the anomaly as early as possible. In this manner, the anomaly can be slowly eliminated from the population.

Another situation is where the dominant gene fails to manifest in a percentage of dogs. Such individuals appear normal, yet they are transmitting the gene to half of their offspring. Pedigrees with this sort of gene behaviour present a picture of erratic heredity, the gene apparently not observing the rules for either dominant or recessive inheritances. Elimination of the anomaly by persistent culling of affected animals is often quite slow.

However, the majority of anomalies are inherited as recessives. The primary reason is that the majority of mutations are due to recessive genes. In addition, recessive genes have the property of remaining in the population as heterozygotes. Hidden by the normal gene (say A), the recessive gene (say *a*) persists in the carrier *Aa*. The breeder may be unaware that some of his animals are *Aa*, until such time as two *Aa* are mated and the recessive *aa* individual makes an appearance. In hindsight, of course, the breeder can classify such parents as carriers. If he breeds from these animals, he does so in the understanding that two-thirds (if they are mated together) or one-half (if they are mated to other animals) of the progeny will be carriers. Persistent selection of abnormal and carrier animals (once detected) will lead to eventual elimination of the gene. In some cases it is possible to detect carrier animals by means of special laboratory tests, in which event elimination is speeded up. In general, however, this is only possible for a minority of cases, although every effort should be made to detect carrier animals.

None of the above indicates how an anomaly can become common in a breed. In fact, the contrary if anything, for slow elimination would seem to be the fate of all genes producing an anomaly. The answer lies in the breeding structure of most dog breeds. If a certain dog is a superlative specimen of the breed, breeders will seek his services for their bitches (if it is a male) or to purchase puppies (if it is male or female). This is a fine arrangment for raising the quality of a breed, for it means that the excellent qualities of the individual are disseminated far and wide via the genes.

However, suppose by chance the individual was heterozygous for a recessive anomaly? The preferential use of the individual at stud or purchase of pups from the stud would distribute the anomaly throughout the breed. The anomaly would not surface immediately but two or more generations later, long after the damage was done, so to speak. In reality it is unlikely that one dog would dominate a breed in this manner, but it is possible for several who have done well in the show ring to do so. This increases the probability that the above scenario would occur, since any one of the dogs could be heterozygous. It is likely that the culprit would be a male dog, since he would sire more offspring than a bitch could hope to mother. No blame can be attached to the owner of the dog, since he or she would have no inkling of what is happening.

The other cause of anomaly is where breeders are selecting for a character, usually an exhibition feature, which causes some individuals obvious distress. Two features which come to mind are the short muzzles of certain breeds which can induce laboured breathing or a prolapsed soft palate. Selective breeding for certain eye shape, or more accurately for eyelid shape, can be inducive of entropion and extropion. Other examples may doubtless spring to mind. Anomalies of this nature can be avoided by not breeding dogs to such demanding standards.

It is unwise to breed from any dog with an inherited anomaly which has been surgically corrected. Even if the defect is relatively minor, it could reappear in later generations in a more severe form. This aspect should not require stressing, but the temptation to breed from such a "corrected" animal can be strong. It should be remembered that while surgical intervention may be desirable to relieve distress for the dog, the gene(s) which produced the anomaly will still be there.

Elimination of Anomalies

Suppose an anomaly has been discovered in a stud, blood-line or breed, what should be done? The answer, of course, is to set in motion plans to eliminate it. One could neuter or destroy all suspected animals, but this is a policy of despair. Apart from the distress this would cause, valuable breeding stock could be lost unnecessarily. True, one must be prepared to accept the birth of animals with the anomaly in the immediate future, but this is the price we must pay for retaining a few suspected individuals which are outstanding in themselves or merit retention for other reasons.

Different procedures must be put into effect, depending upon how the anomaly is inherited. In the case of a dominant, the answer would seem to be straightforward: do not breed from any individual which shows the anomaly. There should be only one permitted exception to this rule. If the anomaly is trivial and the individual is outstanding in other respects, one could breed from it. However, one should do so on the understanding that the anomaly will be passed on to about 50 per cent of the offspring. For a crippling or distressful anomaly, no exceptions should be allowed.

In principle, such a policy could eliminate the anomaly in one generation or within a few generations if it is trivial. However, two complications enter the picture. It is not unusual for dominant anomalies to fail to manifest in a percentage of individuals. This means that an individual may be passing on the dominant gene to its offspring, despite that its is apparently normal. Such individuals can be identified by the appearance of the anomaly among his or her offspring. Initially, there may be confusion as to which is the parent responsible. The mating of the same parents should not be repeated and, if the parents are mated to other dogs, a watch should be kept as to which one produces further anomalous progeny. Once the carrier animal has been identified, he or she should not be bred from again.

The second complication is where the anomaly does not make an appearance until the individual has been breeding for some time. Here, the gene has been passed on before the breeder is aware of what is happening. Once the first signs of the anomaly have been observed, the individual should be withdrawn from breeding. The difficulty is how to deal with the offspring. Doubtless, some will eventually develop the anomaly while others will not. Every effort should be made to identify the first group, even if this means having the dogs examined by a veterinarian specializing in the detection and/or treatment of the anomaly. Early detection can avoid greater heartbreak and problems at a later date.

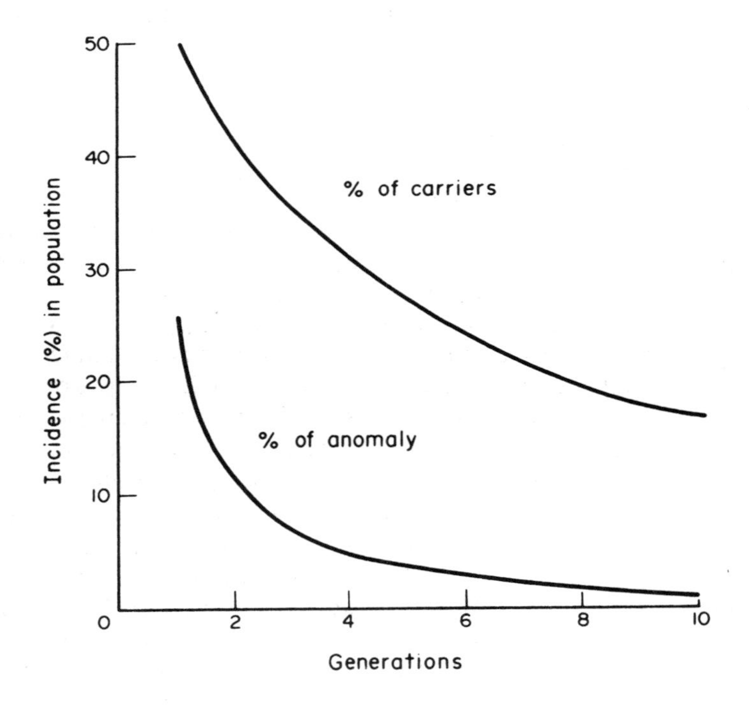

FIG. 32. The expected decline of a recessive anomaly over successive generations when the anomaly is prevented from breeding. Note that, while the percentage of anomalous animals falls quickly, the percentage of carriers does not. This is a primary reason why recessive anomalies appear to persist in populations.

When the anomaly is inherited as a recessive, the problem is a little more complex, but can be tackled by two methods. The first is similar to that suggested above: do not breed from any dog which shows it. The anomaly will not be eliminated immediately, but the incidence will fall quickly at first but then persist at a low level. The progression of events is portrayed by Fig. 32. In generation 1 the anomaly has just been discovered, segregating at 25 per cent; in generation 2, by breeding only from the normal dogs, the incidence has fallen to 11 per cent; by generation 3, to 6 per cent and so on. Only by breeding from wholesome individuals can the incidence of recessive anomalies be kept at a low level, until by chance it disappears completely.

In many studs this steady selection will lead to the elimination of an anomaly, even if only after a number of generations. In the breed as a whole it is not likely to be so successful. The incidence may persist at a low level. This is why some anomalies appear to stubbornly resist final elimination. The reason is the slow decline of carrier individuals. This is shown by the upper curve in Fig. 32. The fall in the incidence of carriers is less rapid and tends to level out at a higher frequency. These carriers are indistinguishable from homozygous normals, but when two of these are accidentally bred together, the anomaly will occur among their progeny.

The final elimination of the anomaly will depend upon the detection of the carrier individuals. When this can be achieved by special tests in a laboratory, the problem is easily solved. Any suspected individual should be tested and if the results indicate that it is a carrier the individual should not be used for breeding, unless there is a compensating reason. The advantages and disadvantages must be carefully weighed. Whatever may be the advantages, the main disadvantage is that 50 per cent of the offspring will be carriers.

The carrier individuals can also be detected genetically. This is accomplished by mating the suspected carrier to an anomalous animal (if the anomaly is trivial) or to a known carrier. If an anomalous pup is produced, then the suspected becomes a known carrier. If only normal pups are bred, there is the possibility that the suspected carrier has produced normal pups by chance. The probability of this decreases as the number of normal pups increases, so the risk of error can be made as small as desired by stipulating that more than a specific number of pups are examined. This concept has previously been discussed in a previous chapter in deciding how many pups should be bred to detect if a black may be a carrrier of blue. The principle involved is precisely the same. The level of permissible error will depend on how careful the breeder wishes to be in not accepting a carrier as a homozygous normal.

To round off methods of test-mating, a general procedure of testing for an anomaly may be described. The method is only practical for males, but does ensure that the dog will not be heterozygous for any anomaly. The

TABLE 15

Number of Pups Required to Attain a Given Percentage of Error in Breeding

No. of pups	A	B	C
1	50	75	88
2	25	56	78
3	12.5	42	71
4	6.3	32	66
5	3.1	24	62
6	1.5	18	59
7	0.8	13	57
8	0.4	10	55
9	0.2	7.5	54
10	0.1	5.6	53
11	–	4.2	52
12	–	3.2	52
13	–	2.4	51
14	–	1.8	51
15	–	1.3	51
16	–	1.0	51
17	–	0.7	50
18	–	0.6	50
19	–	0.4	50
20	–	0.3	50
21	–	0.2	50
22	–	0.2	50
23	–	0.1	50
24	–	0.1	50

procedure is to mate the male to a number of his daughters (which need not be from the same mother). The error involved here is that of not mating the male with a carrier daughter (only half of the daughters will receive the suspected recessive gene) and that of not breeding at least one affected pup. To calculate the error, it is first necessary to find a value for each daughter representing the probability that it is not a carrier. This is shown by column C of Table 15. Then the error for the male is shown by multiplying each of these values together. As an example, suppose a dog has been mated with 5 daughters and the resultant litters consist of 3, 4, 5, 5 and 7 pups, respectively. Consulting column C of the table reveals that the values for the daughters are 71, 66, 62, 62 and 57, respectively. The error for the male is found by multiplying these values together, namely 10 per cent. The values in the table are percentages, so remember to place a decimal point in front of each before performing the calculations.

It has been proposed that the male to be tested should be mated to a number of his sisters. The argument is that if the male has received a recessive deleterious gene from either his sire or dam, then the chances are that one or more of his sisters will have. The chances are indeed 50 per cent. The situation is exactly the same as for mating a male to a succession of his daughters. The same procedure for calculating the chances of error must be followed. Column C is used to find the probability that each sister is not a carrier. The final probability is calculated from these values. This method is not as flexible as that of using daughters, for more daughters can be easily obtained. In fact, daughters bred from daughters can be used if need be. The continuation of the stud or blood-line can go hand in glove with the testing in a sense, except that the testing requires more inbreeding than the breeder may otherwise like to carry out.

The values of Table 15 are for recessive genes with full viability and penetrance. Not all genes for an anomaly will behave in this manner. Some will die before they can be classified, while some individuals will receive the gene yet fail to show it. In these circumstances the values of the table are too optimistic. Whenever such a situation is suspected, the answer is to breed more pups than otherwise would be necessary.

In the case of anomalies due to polygenes, the main remedy is that of selection. Ideally, no individual with the anomaly should be allowed to breed, regardless of how mildly affected it may be. In practice, only the most severely affected animals are culled, the mild cases being passed over, especially if the dog is excellent in other features. This means that the selection against the anomaly is less intense than it should be and the elimination of the anomaly will take longer. An example of selection against a polygenically determined anomaly is that of hip dysplasia. This means that the selection must not falter and that the process may require many generations.

When the anomaly is due to threshold inheritance, the situation is more difficult to handle. The underlying heredity is polygenic, but it is impossible to detect all animals with some of the pertinent polygenes, only those in which the polygenes have actually produced the anomaly. The situation is equivalent to culling the most severely affected animals in straightforward polygenic situations and allowing all of the mildly affected to breed. In appearance all of these will be normal and healthy, regardless of the polygenes they may be carrying for the anomaly. One must be on the alert to detect any individual which has a marked propensity to produce anomalous offspring. These should be ruthlessly culled.

Family selection is more effective than individual selection in dealing with polygenic anomalies. The occurrence of a severely afflicted individual in a litter implies that its sibs may have inherited a dose of the relevant polygenes, even if they are apparently normal. They should also be culled. This is rather drastic, perhaps intolerably so. Nonetheless, if the situation is serious enough, this method may have to be employed. Family selection is the most effective means of dealing with threshold characters.

In practical breeding, it is rarely possible to test-mate females. However, this is not necessarily a drawback if one or more males can be properly tested and pronounced free of a recessive genetic anomaly. No matter how many carrier bitches he may mate, none of the progeny will be anomalous. Some of the carrier bitches will produce carrier offspring, but this cannot be avoided. It does imply that successive stud males must be test-mated for each generation if the anomaly is not to reappear. The testing must continue until such time as the breed is judged to be devoid of carriers.

Some of the tests outlined above may be beyond the capability of individual breeders. In this event, it may be advisable for several people to band together and form a breeders' co-operative as far as the test-mating of certain outstanding males are concerned. If this is impractical, a breed club or society could take matters in hand and arrange a scheme of test-mating. Certified males should command increased stud fees. Adequate supervision by an independent person or body may be necessary to assure correct operation of the scheme.

The recent books of Foley *et al.* (1979) and Hutt (1979) have discussed the nature and incidence of genetic anomalies, the latter somewhat more critically than the former. However, the two accounts do not cover exactly the same ground and are complementary in many respects. The older literature have been thoroughly covered by the monograph of Burns and Fraser (1966). It is fortunate that this summary exists for while much of the older work is interesting and suggestive, it is not particularly reliable genetically. There is in fact little real point in re-reviewing much of the older literature in detail.

Genetically Determined Anomalies

An effort has been made to include all abnormalities for which a genetic basis has been proven. Also others for which a genetic basis seems reasonable but rests upon less secure foundations. There are two reasons why this policy seems desirable. The first is that veterinarians who have been consulted for advice on the anomaly should be aware that the condition could be heritable. Secondly, once the alert has been sounded that the condition could be inherited, an obligation should be felt to collect data which could establish the pattern of inheritance. Anomalies which present signs of being familial should not be ignored. If it is genetic, the anomaly will not go away by the simple expedient of prescribing a new diet or prophylactic medication.

Numerous reports have been (and doubtless will be !) published in which the authors draw attention to a genetic aspect of the anomaly under consideration. It will be appreciated that whether or not one accepts the suggestion will depend upon the weight of evidence. To some, this may be sufficiently conclusive, but to others it may seem nebulous. No account has been taken of such papers unless the evidence is strong. To take account of all such reports would swell the size of this volume out of all proportion to the worth of the information. To say that all anomalies are due in some measure to a genetic factor is to state the obvious. The worthy papers are those which pinpoint the genetic influence and present precise details of the mode of inheritance. Happily there has been a steady increase in the quality of veterinary reports in this respect.

During the last two decades or so, a number of publications have reviewed the literature on genetic anomalies. These have performed the valuable function of garnering the reports which have appeared in a wide variety of genetic and veterinary journals. The more useful of these reviews may be listed as follows: Burns and Fraser (1966), Foley *et al.* (1979), Hutt (1979), Patterson (1980), O'Neill (1981), Desnick *et al.* (1982), Merton (1982), Migaki (1982), Patterson *et al.* (1982), Pidduck (1987) and Willis (1989).

Skin

The hair is a lifeless structure formed from follicles embedded in the outer layers of the skin. It is activity in the skin which determines the character of the coat, its density, fullness, length and the proportions of the different hairs which make up the coat. Hairs, therefore, are appendages of the skin. A previous chapter has described hair length, wire hair and kinky, all of which are due to gene mutation. Since they differ from normal, they are in one sense anomalies, although few people would consider them in this light. These are "legitimate", in that the mutants do not interfere with the well-being of the individual.

Acrodermatitis

As pups, afflicted animals appear to have difficulty in nursing and, when weaned, have difficulty in mastication which may necessitate hand-feeding. By between 5 and 6 weeks, they are noticeably undersized. The legs are held wide apart, with the toes splayed. Skin lesions appear at about 6 to 10 weeks as pustular dermatitis around all of the body orifices, also, particularly on the head, and between the toes. The dermatitis becomes progressively worse, with inflammation and the formation of pus, notably around the eyes and mouth. The ear becomes inflamed and infected, the infection spreading to the lungs. There is diarrhoea of varying severity, from soft faeces to watery bloody excreta.

Normal behaviour may be observed up to about 4 to 8 weeks, when the pup becomes irritable and aggressive, biting litter mates. Later, the animal becomes subdued, plays less frequently and eventually spends most of the time asleep. The average age at death is about 7 months but some individuals have survived for as long as 15 months. The usual cause of death is intractable respiratory infection. The condition is due to a recessive gene *ad*, which was observed in the Bull Terrier (Jezyk *et al.*, 1986).

Alopecia

A remarkable characteristic partial baldness is described in a strain of Miniature Poodle by Selmanowitz *et al.* (1970, 1977 *a*). The loss of hair is closely similar in all affected dogs and is confined to the dome of the head, the rump, legs but not the feet, and most of the stomach. Other areas of the body appear to have more or less normal hair. The condition is produced by a recessive gene *al*.

Black Hair Follicle Dysplasia

This is a remarkable case of partial loss of hair. A number of piebald white-spotted animals showed normal glossy coat for the white areas but a dulling and thinning of the coat for the coloured (black) areas. The anomaly develops about the first month of life. Not all of the black areas are affected. Areas on the side of the head and neck are normal, as are the margins of the black areas with the white. The skin of the black areas are dark and periodically scaly. Entirely white animals appeared normal.

The genetic data are a little unusual. The two original dogs, male and female, both afflicted, produced 15 offspring, all abnormal except for 3 which were all-white. The implication is that they are homozygous for the causative gene. The gene could be dominant or recessive, but the chances are that it is recessive. The gene has been symbolized *bh* (Selmanowitz *et al.*, 1977 *b*).

Blue Dog Syndrome

Several reports have noted that blue dogs of genotype *dd* may have a lustreless thin coat which, in extreme cases, may be associated with variable alopecia. Usually the coat is of normal appearance until puberty. Langebaek (1986) has featured a pedigree of black and blue dogs in which the anomaly recurred in successive generations but was confined to the blue individuals. However, not all of the blue dogs possessed defective coats, as judged by simple inspection. The association between the *d* gene and the anomaly could indicate either a pleiotropic effect of the gene or linkage with a predisposition for the defective coat. The exact cause of the anomaly is unknown but preliminary findings would seem to implicate a hormone involvement.

Connective Tissue Dysplasia

This is a defect of the collagen bundles which are the substantive dermal layers of the skin. The outcome is extreme fragility of the skin and peripheral blood vessels. The skin is loose, has a soft, velvety feel and lacerates easily. Wounds tend to gap, heal slowly and leave wide shallow scars. The affliction is often called the Ehlers–Danlos syndrome after a similar one in man. The condition is due to a dominant gene *Cd* and the above description is for the heterozygote. The homozygote *CdCd* is either identical to the above or is lethal. The data on mating *CdCd* animals together are too meagre to be decisive on this point (Hegreberg *et al.*, 1969).

Dermoid Sinus

These sinuses occur in the Rhodesian Ridgeback either anterior (on the neck) or posterior (towards the tail) to the ridge of reversed hairs. The sinuses are thin tubes of tissue extending from the skin to the spine. They are considered to be due to incomplete separation of the skin from the spine during embryonic development. The sinuses have small external openings which are surrounded by tufts of hair. An individual may have one or several sinuses. In newborns, they can be felt as thin cords under the skin. The sinuses are benign until they become infected, whence they become inflamed cysts requiring surgical excision.

That the sinuses are inherited is not doubted, but there is confusion concerning the manner of heredity. Both dominant and recessive inheritance, as well as polygenic, have been suggested. Much of the confusion arises from the almost certain probability that some dogs which ought to have sinuses because of their genotype do not. This will seriously distort any simple analysis. The association between the occurrence of sinuses and the ridgeback condition prompts the suggestion that the sinuses are an aspect of the ridgeback anomaly. The expression of the ridgeback varies, such as number, position and development of the sinuses. However, dermoid sinuses have been reported in breeds other than the Rhodesian Ridgeback. Thus, the association is not complete, but this does not exclude the possibility that the reversal of hair slope is a predisposing factor.

However, the extensive data of Mann and Stratton (1966) is suggestive that dermoid cysts in the Rhodesian Ridgeback are due to a recessive gene *ds*. There is a small deficiency of affected pups in the data which could be due to impenetrance or to the inability of some breeders (who contributed the observations) to detect very small cysts.

Dermatomyositis

The syndrome develops initially as a severe dermatitis at about 7 to 11 weeks of age, affecting the ear pinnae, around the eyes, lips, tip of the tail and skin over bony protuberances of the limbs. There is partial loss of hair, with inflammation, blistering and crusting of the skin. This is followed by wasting of the muscles of the head, neck, trunk and extremities, resulting in a stiff gait. There is variation of severity but overall consistency of symptoms between individuals. The disease occurred in the Rough Collie and a genetic study indicated a dominant mode of heredity (Haupt *et al.*, 1985; Kundle *et al.*, 1985). The gene is symbolized by *Dmy*.

Hairless

Hairlessness is a frequent mutation, recurring all over the world. In a suitable environment, or one in which the animal is sheltered from extremes of temperature, hairless dogs have flourished, probably as curiosities at the hands of man. Thus it is possible to have the Chinese crested and Mexican hairless dogs, to name the more well known. Given adequate care and attention, these animals are hardy beasts. They are not fully hairless, for the head, feet and tail may display varying quantities of hair. Letard (1930) found that one form of hairlessness was induced by a dominant gene and that the homozygote dies at or soon after birth as a result of defects of the head and oesophagus. The hairlessness of the Chinese Crested dog has also been found to be due to a dominant gene (*Hr*) and with lethality of homozygotes (*HrHr*) (Robinson, 1985). However, the homozygotes in this instance apparently die in the uterus since none was observed.

Two cases of hairlessness born to normal parents have been reported. Zulueta (1949) found three hairless pups among fourteen young born to normal parents. These were hairless except for hair around the muzzle and on the feet. More extensive data have been reported by Sponenberg *et al.* (1988) for the hairlessness shown by the American Hairless terrier. Pups of the breed are born with a thin downy coat which is lost, but the vibrissae hairs persist. The condition is due to a recessive gene symbolized by *ha*.

Conroy *et al.* (1975) has described a partial alopecia in two animals bred from normal coated parents. The coat was of normal appearance until about 5 weeks of age when a progressive thinning could be seen. The presumption is that of recessive inheritance, although this could not be determined.

Lentiginosis Profusa

The anomaly consists of numerous black spots of varying diameter on the skin of the lower legs and stomach. These appeared at any time between one and four years of age. After persisting for an unspecified period, the spots began to diminish in size and number. The lesions were observed in the Pug (Briggs, 1985).

Sensory Organs

Of the sense organs, keen eyesight is of importance, especially for working dogs. The eye is a complex organ, with many parts which can go wrong, and it is readily accessible for detailed examination. Possibly for all of these reasons the eye has the greater number of anomalies discussed in this section. Several have been of very real concern to breeders and have occupied the attention of many veterinarians seeking to understand the aetiology and inheritance of the various anomalies.

Aphakia

Aphakia is the congenital absence of the lens of the eye. The defect has been reported in a family of St. Bernard dogs. However, although the evidence indicates monogenic recessive inheritance, this is too limited in amount to be conclusive (Martin and Leipold, 1974).

Cataract

Cataract is defined as an opacity of the lens of the eye, usually apparent by naked eye inspection as dull whiteness of the pupil. The cataract is said to be primary when it is unaccompanied by any other anomaly of the eye. If it is accompanied by another anomaly, it is said to be secondary. The distinction is important because cataract is a common feature of several other eye defects and is merely a consequence of the other, usually more serious, defect. This section is concerned with primary cataracts.

There are many forms of cataract, and distinctions between these can only be achieved by examination of the lens by special instruments. Most cataracts develop slowly and the age of onset varies between cataracts and breeds. Most are instrumentally detectable long before they become visible to the naked eye. It is unusual for the cataract to be other than bilateral, although the rate of development in each eye may differ.

Some of the early claims for inheritable cataracts rest upon inadequate data for the mode of inheritance to be properly determined. The approach to the problem of mode of inheritance has now become more critical and the more recent researchers have made commendable efforts to obtain decisive data. Table 16 summarizes the most thoroughly investigated cases of cataract occurring in different breeds. No stigma should be attached to the citation of breeds, these being mentioned for identification purposes, because cataract may occur in any breed at any time. It should be noted that different cataracts may be either recessive or dominant in heredity. This is an indication that different genes are concerned. It is impossible to assess how many independent genes are involved. In principle, each case represents a new mutant allele until proven otherwise. The more informative reviews of cataract in the dog are Barnett (1976, 1978, 1985b, 1988), Curtis (1982) and Curtis and Barnett (1989).

TABLE 16

The Major Cases of Cataract

Breed	Age of onset	Nature of cataract	Inheritance	Reference
Afghan Hound	1–3 yr	Anterior and posterior subcapsular. Progressive.	Probably recessive	[12]
American Cocker Spaniel	6 months	Outer nuclear area. Usually progressive.	Probably dominant	[17]
American Cocker Spaniel	2–4 yr	Anterior or posterior cortical. Progressive.	Probably recessive	[16]
Boston Terrier	8 wk	Subcapsular. Progressive.	Recessive	[1]
Boston Terrier	2–3 yr	Radial. Slow progression.	Suspected genetic	[6]
Chesapeake Retriever	--	Suture, equatorial.	Dominant	[7]
German Shepherd Dog	Birth	Lamellar. Little progression.	Dominant	[4,9]
German Shepherd Dog	8–12 wk	Central-posterior. Progressive.	Recessive	[9]
Golden Labrador Retrievers	6–12 months	Polar or peripheral. Latter is progressive.	Probably dominant	[1,13]
Miniature Schnauzer	Under 1 yr	Posterior subcapsular.	Recessive	[15]
Miniature Schnauzer	Birth	Nuclear.	Recessive	[3,8]
English Sheepdog	1–2 yr	Cortical to total. Progressive.	Probably recessive	[10]
Staffordshire Bull Terrier	Under 1 yr	Subcapsular. Progressive.	Recessive	[1]
Standard Poodle	Under 1 yr	Equatorial. Progressive.	Recessive	[5,14]

Welsh Springer Spaniel	Few months	Central. Progressive.	Recessive	[2]
West Highland white Terrier	Under 6 months	Posterior "Y" suture.	Recessive	[11]

References: [1] Barnett, 1978; [2] Barnett, 1980; [3] Barnett, 1985a; [4] Barnett, 1986; [5] Barnett and Startup, 1985; [6] Curtis, 1984; [7] Gelatt *et al.*, 1979; [8] Gelatt *et al.*, 1983; [9] Hippel, 1930; [10] Koch, 1972; [11] Narfstrom, 1981; [12] Roberts and Helper, 1972; [13] Rubin, 1974; [14] Rubin and Flowers, 1972; [15] Rubin *et al.*, 1969; [16] Yakeley, 1978; [17] Yakely *et al.*, 1971.

Collie Eye Anomaly

This eye defect is (or was) common in the Collie (including the Shetland Sheepdog), hence the name given to it. The defect is not known to occur to any extent in other breeds. Collie eye anomaly (CEA) must be treated as serious because of its high frequency of occurrence. Surveys of Collies have shown that as many as 80-90 per cent of dogs had it. These high percentages are now falling due to selection programmes aimed at reducing the incidence. However, the incidence is still high enough to cause concern. One report has cited a reduction of CEA from 97 to 59 per cent over a 3-year period (Yakely, 1972).

CEA involved a major disruption of normal development of the back of the eye. The retina, choroid and sclera layers, retina blood vessels and optic disc are all anomalous. As many as eight different anomalies have been defined as manifestations of CEA. These may vary in severity, and affected dogs need not have all of them. The severity of CEA varies from unusual tortuosity of the retina blood vessels, through graded anomalies of the retina and choroid, to retina detachment and haemorrhaging. CEA is rarely progressive, except for retina detachment which may be precipitated as a consequence of other anomalies. All of the defects may be seen under the ophthalmoscope. Only the more severe cases are blind and in most instances this is the first indication that a breeder may have that his dogs may have CEA. However, routine ophthalmoscopic examination by an experienced person can detect CEA in puppies at 6 weeks of age. The anomaly is apparent in both eyes, but not necessary to the same degree.

Breeding experiments and pedigree analysis have shown that CEA is produced by a recessive gene *cea* (Yakely, 1968; Bedford, 1982a). However, it is evident that the gene has many diverse effects (i.e. is pleiotropic), even if these affect the same or related eye structures. At one time it was thought that a selection programme might be successful if only the more severely affected dogs were prevented from breeding, but this has been largely abandoned in favour of the view that all grades of the affliction should be treated equally. Dogs with minor defects are perfectly capable of breeding pups with the most severe defects. There is an extensive literature on the disease and this has been reviewed by Bedford (1982a). The defect is not confined to the Rough and Smooth Collie but also occurs in the Shetland Sheepdog.

Hutt (1979) does not agree that CEA is due to a single recessive gene. He proposes that the condition is polygenic, citing in favour of this view the high incidence of CEA and the wide range of expression of the defect. He also criticizes the interpretation of some of the results purporting to establish monogenic heredity. Although Hutt's comments may raise some doubts upon current thinking, they do not amount to total refutation. They do, however, encourage a critical appraisal of the evidence and should prompt the collection of additional data which may be decisive in settling the problem.

Hutt picks up an earlier suggestion that the remarkably high incidence of CEA in the Collie might result from breeders favouring an attribute which is conducive to it. He makes the point that the skull of the Collie is extremely long and narrow. The most common anomalies of CEA are those to be expected if the eyeball is exposed to unusual pressure from behind and outside it. Could the narrow head be distorting the size and shape of the eye socket, thus interfering with normal eye development? An objection to the suggestion is that other breeds with elongated skulls are not plagued with CEA.

The Collie breeds have the merle gene *M* as part of their genetic constitution, and the gene has been implicated in CEA. The *M* gene does indeed produce eye defects when homozygous, and may interact with CEA to produce more severe effects. However, the *M* gene is independent of the postulated *cea* gene. This is established by the presence of CEA in non-merle dogs.

Crystalline corneal dystrophy

The anomaly appears as a hazy oval or spherical opacity in the cornea of the eye. The onset of the defect is variable, appearing as early as at five months in some animals but being delayed for as long as one or two years in others. Fortunately, the opacity does not become sufficiently dense to interfere noticeably with the vision. Genetic studies revealed that the opacity is inherited in a recessive manner (Waring *et al.*, 1986). The gene was discovered in the Siberian Husky and is symbolized by *cr*.

Dobermann Eye Anomaly

The eyes are seriously deformed with this recessive congenital anomaly. The eyes of afflicted pups open a little later than their normal-eyed litter mates, between 14 and 17 days compared to 12 to 13 days. The skull appears to be unusually narrow, while the eyelids are smaller and the eye bulb is reduced in size. A striking feature is the dull whiteness of the opaque cornea and a prominent third eyelid. At autopsy, the whole eye was found to be seriously distorted. The anterior chamber was absent, with neither a pupil nor lens. The retina was detached and the retinal epithelial cells were partially depigmented. The anomaly has been reported for the Dobermann in Norway and in the United States (Bergsjo *et al.*, 1984). The gene has been symbolized as *dea*.

Ectropion and Entropion

Ectropion is the eversion or turning outward of the eyelid and entropion is the inversion or turning inward of the eyelid. These are common complaints of dogs and they illustrate how specific characters may be influenced by others of a more general nature. The common factor in many cases is excessive facial skin of many breeds which can cause either defect. For example, the loose skin may cause the lower eyelid to fall away from the eyeball, inducing ectropion. Entropion may be due to too small an eyelid opening or when the eyelids are too slanting. The diamond-shaped eye, favoured in some breeds, is a constant source of trouble in this respect. Either defect can lead to excessive weeping, irritation, inflammation and serious eye damage if neglected (Barnett, 1976).

Both ectropion and entropion are inherited in the sense that the degree of "looseness" or "tightness" of the facial skin or shape of the eyelids are polygenically controlled. Whether there are independent genes for either defect is debatable. Either defect may be familial but be inherited irregularly because the primary cause is not being studied. Particular attention should be given to the role of facial skin or eyelid shape in any attempt genetically to combat cases of eyelid deformities.

Everted Third Eyelid

The usual symptom of this defect is a running eye, which may be watery or be of a muscoid nature. The anomaly takes the form of a tight curling forward of the membrane nictitans in the corner of the eye. Either one or both eyes may be affected, and the eversion is not always readily apparent. The incidence is familial and probably due to recessive gene, although more data are required if this latter supposition is to be taken seriously (Martin and Leach, 1970).

Glaucoma

Glaucoma is an increase of pressure within the eye bulb which, if left untreated, can lead to serious physiological disruption of function and blindness. Two categories of glaucoma are recognized; primary, where it is unaccompanied by any other ocular defect, and secondary, where it is accompanied by other ocular defects. In the latter, the glaucoma may be the main cause of the defect or be a consequence. This section is concerned with primary glaucoma.

Primary glaucoma arises when an imbalance occurs in the production of aqueous humour and drainage. Most of the drainage occurs via the iridocorneal angle and the ciliary cleft. In one form of glaucoma, the iridocorneal angle is abnormally narrow and drainage is impaired. In another form, the angle is normal but the outflow of aqueous humour is less than normal due to causes not precisely known. The two forms are known as angle-closure and open-angle glaucoma, respectively.

Glaucoma is not one of the more common eye defects but, nevertheless, it has been reported for a number of breeds. Gelatt and Gum (1981), for example, list as many as 16 breeds in which it is known to occur. The malady is usually assumed to be inherited as a recessive but scarcely any evidence is available to substantiate the assumption.

On the other hand, Gelatt and Gum (1981) have shown conclusively that primary glaucoma is inherited as a recessive (symbol *gl*) in the Beagle. The typical age of onset was between 9 and 18 months old, as indicated by the raising of intraocular pressure. The iridocorneal angle slowly closes by 2 to 4 years and the eye can become extensively damaged. However, the manifestation of similar glaucomas in other breeds is stated to be later than that observed in the Beagle: some 3 to 4 years for the Norwegian Elkhound, 6 to 7 years for the Keeshound, and over 8 years for the Miniature Poodle (Bedford, 1980). Cottrell and Barnett (1988) have reported on the malady in the Welsh Springer Spaniel, in which it is probably inherited as a dominant, with an average age of onset of between 3 and 4 years but with much variation in this respect. Also, there is variation in expression of the iridocorneal angle even in clinically normal dogs, a feature which the authors speculate may be representative of some of the heterozygotes.

Hemeralopia

Hemeralopia or "day blindness" is an inability to see in bright light, although vision in dim light or at night is not impaired. The defect manifests about 8 weeks of age. Affected animals may collide with obstacles and miss steps which are negotiated with ease in dim illumination. By the aid of electroretinography, affected pups may be positively indentified by 6 weeks. The condition is due to a recessive gene *he* and has been observed in the Alaskan Malamute and Miniature Poodle (Rubin *et al.*, 1967).

Lens Luxation

This is a term to denote the displacement of the lens from its normal position. The reason is a weakening of the ligaments holding the lens. It is a defect which occurs in adult dogs aged between 3-5 years. Most, but certainly not all, cases are bilateral, although there may be an interval before both eyes are affected. The luxation is often accompanied by secondary glaucoma. The luxation produces blindness as revealed by the animal blundering into obstacles. The pupil may be dilated and "staring". The dog may rub its eye as if the luxation is causing distress. An indication that the condition is inherited is shown by its prevalence in certain breeds, notably the Terrier group (Barnett, 1976). Willis *et al.* (1979), Curtis and Barnett (1980), and Garmer (1986) have shown that the anomaly is due to a recessive gene *lx* in the Tibetan Terrier.

Curtis *et al.* (1983) have described a similar recessively inherited defect of the lens in the Miniature Bull Terrier which could well be produced by the same recessive gene. A mating between affected individuals of the two breeds gave birth to two offspring which, at 6 months of age, showed degenerative changes to the eye of a nature to be expected for the anomaly.

Microphthalmia

Gelatt *et al.* (1981) have tabulated breeding data which suggested that, in the Australian Shepherd dog, microphthalmia is inherited as a recessive but with a small degree of impenetrance. However, the incidence of the anomaly was closely associated with the assortment of the merle (*M*) gene. Unfortunately, only a limited number of offspring were obtained from those matings which could have revealed the independent assortment of merle and the depicted microphthalmic gene. The possibility exists that the eye anomaly is a pleiotropic affect of the *M* gene. The small eyes of the microphthalmia were merely an obvious outward sign of anomaly since many affected individuals had other ocular defects, such as cataract, retina degeneration and detachment.

Microphthalmia is the most obvious feature of the eye anomalies observed by Laratta *et al.* (1985) in the Akita breed. In addition to the small eye aperture, there were congenital cataract, bulging of the lens and retina dysplasia. A genetic causation seems feasible but the mode of inheritance could not be determined.

Multifocal Retinal Dysplasia

This is a developmental defect of the tapetum, retina and optic nerve. These may be seen ophthalmoscopically in puppies as soon as 3-4 weeks of age. Small sections of the tapetum show decreased amounts of pigmentation. The retina displays many small areas of deformity, such as folding and rosette formations, as well as pigment deficiency. These do not cause blindness nor other apparent visual disturbance to the affected dog. The defect is attributed to a recessive gene *mrd*, observed in the American Cocker Spaniel (MacMillan and Lipton, 1978).

Persistent Pupillary Membrane

The pupil of the eye in the foetus is covered by membrane which disappears during the first 3-5 weeks of life. However, the membrane may persist in some individuals (hence the name); persisting strands may cross the pupil and be responsible for small cataracts. There is usually no effect on vision in uncomplicated cases but it can be associated with more serious defects, such as cataract and lens anomalies.

The malady has been noted in some ten breeds and is considered to be inherited in some of these. However, there is no firm evidence for the mode of inheritance for any of the breeds. On the other hand, there is reason for thinking that a genetic basis is quite probable for the anomaly in the Dobermann Pinscher and particularly in the Staffordshire Bull Terrier (Curtis *et al.*, 1984; Leon *et al.*, 1986).

Progressive Retinal Atrophy (Peripheral)

One of the scourges of dog breeding of the last few decades has been the problem of progressive retina atrophy (PRA). This is a defect of the eye, where the visual layer of the retina progressively degenerates. There is more than one form of PRA, the most common being the peripheral or generalized PRA. This is so-named because the degeneration commences at the edges of the retina and progresses inwards as the animal becomes more and more severely affected. The first signs are a dimness of vision at night (the dog becomes wary of going out at night), then in dim light and in daylight; finally the animal is totally blind. The process is slow and may take months or years. The pupil of the eye becomes dilated and the pupillary contraction in response to bright light is reduced. A cataract may eventually form. Both eyes are involved.

One of the main aspects which made PRA such a scourge is that the dogs with the defect may be undetected until some time after they have reproduced. The age of onset of clinical signs varies between breeds. The age at which PRA commences to manifest is variable; in the Long haired Dachshund about 12 months or less, in the Norwegian Elkhound by about 2 years, in the Cocker Spaniel at several years and in the Miniature and Toy Poodle as late as 6-9 years; there being much variation within breeds for individual animals. PRA has, in fact, been detected in many breeds and few breeds can be held up as being completely free of the defect. On the other hand, the incidence is low for most breeds.

In every breed in which the inheritance of PRA has been adequately analysed, recessive heredity is apparent. The simplest assumption is that the same recessive gene *pra* is involved in each breed. This is a relatively safe assumption for related breeds that may have common ancestors in the remote past. However, the assumption is less safe for unrelated breeds.

Here, the mutant gene may have occurred independently either at the same or at different loci. Barnett (1976) makes the point that the condition shows remarkable clinical similarities for all affected breeds. This is one reason for postulating the same gene for PRA for all breeds. On the other hand, there are the obvious differences between breeds in the age of onset. This could be due to breed differences in the development of the defect, this being by no means improbable. Future studies may elucidate this dilemma.

One of the problems of diagnosis of PRA is the late and variable onset of the disease. Dogs examined too early in life may not possess clinical signs of the defect and be passed as "normal". However, the defect may be detected by electroretinograms of the response of the eye to flashes of light. These can pick out dogs with insipient PRA long before clinical signs begin to appear. For example, in the Norwegian Elkhound as early as 6 weeks and in Miniature Poodles by 10 weeks. It is obvious that this technique would be of great value in detecting homozygous PRA animals before any breeding is undertaken, much earlier detection of heterozygotes by more rapid assessment of their young, and possibly allow test-matings to be performed which would otherwise be impracticable because of the time involved (Aguirre and Rubin, 1972; Preister, 1974; Barnett, 1976; Wolf *et al.*, 1978; Dice, 1980; Bedford, 1984*a*, 1989).

Aguirre *et al.* (1978) have found that the PRA in Irish Setters may be traced to a deficiency of cyclic GMP phosphodiesterase activity. This is a vital enzyme in the normal metabolism of the retina cells and an impairment of function could lead to cell degeneration. For those interested in the fundamentals of PRA maldevelopment, this is an interesting discovery.

Progressive Retinal Atrophy (Central)

The second form of PRA is known as central progressive retinal atrophy (CPRA) and differs from that described earlier. There is no sign of night blindness, indeed affected animals often seem to be able to see better in dim rather than in bright light. They do, however, have difficulty in avoiding obstacles and may even blunder into a person who is standing still. Under the ophthalmoscope, scattered spots of brown pigment may be seen on the tapetum (the retinal layer responsible for "eye shine"). These spots are not observed in the peripheral PRA. Degeneration occurs in the central area of the retina, hence the veterinary name for the malady. The defect becomes progressively worse until the individual is totally blind. There is dilation of the pupil, but not until the later stages, and cataracts are uncommon.

The onset of the defect is subject to considerable variation, ranging from 2 to 10 years, the majority of cases occurring between3-6 years. Ophthalmoscopically, retinal changes may be recognized as early as between 18 months and 2 years. Because the dog retains peripheral vision, the defect is often not noticed until several years later. Pet dogs, for instance, may not be diagnosed until as late as 10 years, when the owner may consider the blindness to be due to old age.

The distribution of CPRA in breeds is not the same as for PRA. The former occurs mostly in working dogs, the Collies and Retrievers having the highest incidence. In essence, CPRA is not so widespread as PRA. The exact mode of inheritance is stated to be uncertain, but the evidence points towards a dominant gene, but with some degree of impenetrance (Barnett, 1969). A lack of manifestation in some individuals is not uncommon for genes with a wide range of expression.

Recently, Bedford and Willis have found evidence that CPRA in the Briard dog is inherited as an autosomal recessive (Willis, 1989). This could indicate that the condition is genetically different from that discussed above or that the Briard breed genotype modifies the expression towards recessivity.

Rod–cone dysplasia

A progressive retinal atrophy has been studied in the Collie under the designation of rod–cone dysplasia and has been determined to be inherited as a recessive (Wolf *et al.*, 1978). In this respect, the defect behaves in an identical manner to that found in other breeds. However, Wolf *et al.* (1979) have mated an affected Collie with an affected Irish Setter, a union which produced six pups with apparently normal eyes. This is evidence that the condition is due to independent genes, in spite of the closely similar clinical symptoms for PRA in the two breeds. If the symbol *pra* is retained for the PRA gene in the Irish Setter (and probably other breeds), the symbol *ro* may be assigned to the gene in the Collie.

Retinal Dysplasia

This condition is probably congenital, but becomes noticeable at about 3-6 weeks of age. The afflicted pup behaves as if blind: colliding with objects, is less lively than usual and may have a vacant expression. The blindness is confirmed by diagnosis of a detached retina. The condition has been observed in the Bedlingham Terrier, Labrador Retriever, Sealyham Terrier and Yorkshire Terrier, and is due to a recessive gene *rd*. It is interesting that retinal dysplasia (RD) is one of those defects which is traceable (by pedigree analysis) to a particular stud dog (in the Labrador Retriever), where it probably arose as a mutation (Rubin, 1963, 1968; Ashton *et al.*, 1968; Barnett *et al.*, 1970; Barnett, 1976; States, 1978; Bedford, 1982*b*, 1984*b*). The caveat must be issued that the same gene for RD may not be involved in each breed.

Tapetal degeneration

The tapetum is a layer of pigmented cells at the back of the eye and functions to enhance vision in dim illumination. The tapetum is responsible for the reflective "eye shine" which can be seen in the canine or feline eye when a torch is focused on the eye from a distance. An anomaly of the tapetum has been reported in a strain of Beagles, inherited as an autosomal recessive (*td*). The tapetal layer of cells is normal until about 21 days of age, whence progressively degenerative changes may be observed. By between one and two years of age, the tapetum has completely disappeared. The retina appears to be morthologically normal, with merely a small impairment of function. However, there is general deficiency of pigmentation of the eye, with some structures having immature melanosomes and most others showing a patchy distribution of melanin (Bellhorn *et al.*, 1975; Burns, 1988*a*, 1988*b*). It is possible that night vision could be adversely affected.

Merle

The merle gene has been discussed elsewhere. The gene produces a blue eye, either unilaterally or bilaterally, and an absence of the tapetum, the pigmented layer which gives rise to "eye shine" reflection in the dark in heterozygotes *Mm*. The homozygote *MM* is more severely affected. The coat is completely white, the eyes are blue and frequently reduced in size (microphthalmia) or occasionally absent (anophthalmia). The whole eye, in fact, is anomalous (Lucas, 1954; Dausch *et al.*, 1977; Klinckmann *et al.*, 1986, 1987; Herrmann and Wegner, 1988). It has even been suggested that merle of both genotypes have a predisposition to glaucoma due to elevated intraocular pressure (Klinckmann and Wegner, 1987). The majority of animals have hearing impairment or are totally deaf (Mitchell, 1935; Reetz *et al.*, 1977). Somewhat curiously, there is interference with sperm formation (Treu *et al.*, 1976).

Deafness

Deafness, either partial or complete, occurs rarely in dogs, but the affliction has been known to recur in the Dalmatian. The affliction has been studied by Anderson *et al.* (1968), who, however, could arrive at no firm conclusion for the mode of inheritance. If the condition is monogenic, then the gene shows impenetrability and there is little indication of dominance or recessiveness. The complication is that some dogs were only slightly impaired and such animals could easily pass undetected and yet hand on a greater degree of deafness to their offspring. If anything, deafness could be a polygenic threshold character, the precision of the analysis being dependent upon the quality of the detection equipment. The cause of the deafness is a collapse of the cochlea and degeneration of the organs of Corti, both of which are vital for sound hearing.

Deafness is part of the merle syndrome and many of the white merles have some degree of hearing up to complete deafness. A proportion of heterozygous merle also have hearing imperfections (Mitchell, 1935; Reetz *et al.*, 1977). It is commonly believed that white Bull Terriers may be deaf and it would seem that a proportion are, but deafness was also observed among coloured Bull Terriers, hence the heredity of the affliction would seem not to be confined to the white coat (Hirschfeld, 1956). An association between extreme white coat and deafness is not to be unexpected since the embryonic disruption of tissues which induces the white coat often—though not always—disturbs those destined to form part of the inner ear.

Skeleton

The skeleton, as represented by bone structure, is a vital part of the body, governing the size and stance of the dog. It is small wonder that the skeleton is subject to considerable genetic modification. Quite a few of the modifications have been brought about by man in his pursuit for distinctive breeds. Many others, on the other hand, have been produced by mutations which are decidedly unwelcomed. These have resulted in crippling pathological conditions.

Achondroplasia

Achondroplasia, or "disproportion dwarfism", is where the legs are shortened but the head and body remain of normal size. The Basset Hound and Dachshund are examples of achondroplasic animals. This does not necessarily mean that these dogs suffer from a disease; the term merely means that the long bones of the leg do not attain normal length. There is, in fact, some debate whether or not the Basset or Dachshund possess a major gene for short legs. Stockard (1941) found that a semi-dominant gene for short legs assorted from initial crosses between a Basset and German Shepherd Dog. However, Whitney (1948) could not confirm these findings for initial crosses between a Basset and Bloodhound. He found blending or polygenic inheritance.

One point may be stressed, however, speaking more generally. Whitney states that in his experience in crosses between breeds with different leg lengths the shorter length predominates. That is the shorter of the two leg lengths is likely to be obtained in crosses and long-legged dogs are more likely to be bred from short-legged parents than vice versa. The implication is that, while differences in leg length are polygenic, those for the shorter length are dominant to those for the longer length.

Although polygenic heredity may be assumed for length for most breed differences, this does not mean that major achondroplasic genes do not exist. On the contrary, many such genes are known in mammals. Typically, these are to be found assorting within breeds rather than in crosses between breeds. Whitney (1947) described such a recessive gene *ac* in the Cocker Spaniel. The short-legged animals were distinctly different from the normal Cocker. Similar genes have been found by Gardner (1959) in the Poodle, by Whitbread *et al.* (1983) and Lavelle (1984) in the English Pointer and by Meyers *et al.* (1983) in the Samoyed. In all of these breeds, the dwarfing genes are inherited as recessives. It is unlikely that the same gene is involved since, in each case, the short-legged condition was associated with additional but different anomalies.

Bithoracic Ectromelia

Pups with missing limbs occur at a very low frequency in litters and are usually attributed to mishaps of development. As such, they are not normally considered to be inherited. However, a remarkable case of congenital absence of the forelegs has been described by Ladrat *et al.* (1969) who were able to show that it was caused by a recessive gene *be*. A female was observed without front limbs except for vestigial humeri. The hind legs were normal and the animal was able to walk upright upon them. The dog, a bitch, bore pups which were identical to herself. There was considerable mortality among the ectromelic pups.

An interesting instance of partial absence of the forelegs has been described in the Chihuahua by Alonso *et al.* (1982). Only four animals were examined but these were bred from two sets of normal parents, both of which were half brother and sister. The evidence for recessive inheritance is slight but certainly suggestive.

Brachydactyly

Green (1957) briefly mentions the existence of a recessive gene reducing the size of the outer toes of the forelegs and sometimes those of the hindlegs in addition. No other information is available. The gene may be provisionally symbolized as *bd*.

Carpal Subluxation

This defect involves a ventral displacement of the carpal or wrist bones, so that eventually the animal is actually walking on the portion of the foot from the wrist to the toes of the front feet. Afflicted pups are normal until about 3 weeks, when the pups commence to walk, becoming steadily worse. The final degree of displacement is variable and is always bilateral. The condition is inherited as a sex-linked recessive character, determined by a gene *cs* borne by the *X* chromosome (Pick *et al.*, 1967).

Cervical Spondylopathy

The fundamental deformity is a narrowing of the vertebral canal. This puts pressure on the spinal cord, resulting in ataxia and a staggering wobbly gait, the latter deteriorating into paralysis within a few months. The condition has been observed in the Basset Hound (Palmer and Wallace, 1967) and in the Dobermann (Mason, 1977) with a suspected genetic aetiology. Subsequently, Jaggy *et al.* (1988) have shown that the anomaly is inherited as a recessive in the Borzoi. The gene has been symbolized by *csp.* The anomaly also has been termed the wobbler syndrome.

Cleft Palate and Hare Lip

These are two anomalies which occur at a low frequency in most breeds. In most cases, sporadic instances are due merely to accidents of development at the embryo stage. However, if a number of cases recur in the same strain or breed, a genetic basis should be suspected. Wriedt's (1925) data on the heredity of cleft palate in the Bulldog and those of Turba and Willer (1997) in the German Boxer are certainly suggestive of the action of a recessive gene *cp.* However, the data of Weber (1959) for the Bernese Sennenhund are suggestive of dominant inheritance but are too imprecise to be readily accepted.

Despite the above results, it is probably wise to be cautious in accepting single gene determination of either of these anomalies. It is possible that they are polygenically determined threshold characters. It is possible for threshold characters to simulate monogenic inheritance in some matings but give different results in others. The two defects discussed here are physiologically related. Either one can occur independently of the other, yet it is common for them to occur together, although one may predominate in frequency.

Craniomandibular Osteopathy

Typically this is an anomaly of the lower jaw, consisting of the formation of excessively dense bone which interferes with normal mastication. Other bones of the skull and legs may show a similar condition. The CMO defect occurs most frequently in the West Highland White Terrier and in the Scottish Terrier. Typically, the malady becomes noticeable between 4 and 7 months of age by signs of discomfort with chewing and gnawing. A regimen of corticosteroid medication is usually successful and the jaw assumes a normal appearance. The temptation is to breed from such animals, thus perpetuating the anomaly. The work of Padgett and Mostovsky (1986) has shown that the anomaly is due to a recessive gene which may be symbolized as *cmo*. The defect is sometimes called Scotty or Westy jaw.

Dew Claws

The heredity of extra toes on the feet is not straightforward. It is a feature which has been examined on several occasions without yielding clear-cut results. Dominant inheritance is indicated, but with impenetrance. That is, some dogs which ought to have dew claws, according to their offspring, do not have them (Keeler and Trimble, 1938). Whitney (1947) strongly supports the concept that dew claws are due to a dominant gene, but gives no supporting evidence. An alternative explanation is that the presence of extra toes is a threshold character. This would go some way to explaining one account in which dew claws behaved as a recessive (Burns and Fraser, 1966).

Dwarf-anaemia

This dwarfism was observed in the Alaskan Malamute. Afflicted individuals have shorter legs than normal, with enlarged carpal joints. The front legs are bowed inwards as if due to rickets. The skeletal defect appears to be a rise from a thickening of the growth plate, accompanied by an impairment of conversion of cartilage to bone. There is a mild macrocytic haemolytic anaemia. The red blood cells display a number of anomalies while the white blood cells are unaffected. The two conditions are induced by a recessive gene *dan*. It is possible to detect heterozygotes by a careful haematological analysis (Fletch *et al.*, 1975; Sande, *et al.*, 1982).

Dwarf, retina dysplasia

This form of short-legged dwarfism is associated with a variety of ocular anomalies. These range from a mild defect, such as cataract, to extensive maldevelopment of the eyes which can involve retina dysplasia, folding and detachment. The syndrome of defects shows monogenic inheritance but with differential expression between the limb and eye anomalies. The homozygote is dwarf, with defective eyes, while the heterozygotes are of normal size but may display some of the milder eye lesions. The gene is formally designated as a recessive with the symbol *drd* and was observed in the Labrador Retriever (Carrig *et al.*, 1977, 1988).

Elbow Dysplasia

This is one of several joint anomalies which result in lameness. It is a defect of the growth processes of the elbow joint of the forelegs and usually occurs at about 4–6 months of age. Either one or both legs may be affected. Surgical correction is relatively easy and effective and, possibly for this reason, the defect is not always viewed as seriously as it should be. It is apparent that this procedure, although necessary for any afflicted animal, does not reach the genetic liability to the defect. The genetic basis to the defect is that of a polygenic threshold character (Corley *et al.*, 1968; Hutt, 1979). The disease referred to as elbow arthrosis by Grondalen and Lingaas (1988) is an aspect of the lameness. These authors found that the affliction is due in part to polygenic heredity. The section on hip dysplasia may be profitably read in conjunction with this one.

Elbow Subluxation

The main symptom of this defect is a lameness of one or both forelegs, usually by 3–4 months of age. When walking, the animal tends to swing its elbows sideways, with almost all of the movement carried out by the shoulders. The elbows cannot be fully extended and some dogs exhibit signs of pain upon manipulation. The defect is attributed to premature closure of the distal ulnar physis. Although the data are meagre, these are consistent with the assortment of a recessive gene *es* (Lau, 1977).

Hip Dysplasia

Hip Dysplasia (HD) is a serious defect because it can be crippling and because it is (or was) so prevalent (Priester and Mulvihill, 1972). It is a defect of the hip joint, more specifically a failure of the head of the femur to fit properly into the acetabulum (socket of the hip bone). If the fit is not satisfactory, the femoral head may be loose, resulting in excessive rubbing, dislocation and eventual osteoarthritis. The defect may occur in any breed, but the larger breeds are the most affected. It is most congenital, but may develop at any time up to 18 months, but usually sooner than this in most individuals. The defect may be diagnosed much earlier by radiography of the pelvic region. The outward signs are unwillingness to exercise, lameness, stiffness of the hip joint and, later, a wasting away of the hip muscles. Both sexes are equally affected.

There has been considerable discussion of the factors involved in the predisposition to hip dysplasia, a fact which highlights the complexity of the problem. The main factors are: shallowness of the acetabulum, imperfect formation of the femoral head (either of these defects will lead to looseness of the hip joint), the downward slope of the back from the shoulders to the hips, and the size and strength of the muscles controlling femur movement (muscular mass, as it has been termed). Because the defect often manifests during the period of rapid growth, even this aspect has been implicated. The concept is that too rapid growth elicits hip dysplasia, not necessarily in all animals but in those which have the predisposition. The suggested remedy is to slow down the rate of growth!

Initially, it was thought that hip dysplasia was due to a dominant gene with incomplete penetrance, but this seems to be incorrect. It is now assumed that the defect is mediated by polygenes. It could be termed a threshold character if one simply classifies dogs as normal or afflicted. However, this is an oversimplification, for the severity of the affliction can vary, even to simple clinical inspection. The variation can be made more precise by radiographic examination. This can reveal the extent of the dysplasia more or less directly. The extent can be graded, usually as follows: (1) normal or excellent, (2) fair, (3) mild, (4) moderate and (5) severe. Although these grades or scores are arbitrary and the variation only divided into five grades, they nonetheless illustrate the variable aspects of the defect.

Evidence for the genetic aspects may be adduced from three sources. The first is simple recording of offspring from normal or affected parents as normal and afflicted, noting the proportions for the three possible matings. Thus, in a sample of 1154 offspring, those from normal parents showed an incidence of 38 per cent afflicted, those with one parent affected showed an incidence of 45 per cent while those with both parents affected showed an incidence of 84 per cent. The genetic influence, as shown by the increase in per cent afflicted according to whether one or both parents are affected, is unmistakable. A similar picture emerges from Table 17, which takes into account the severity of the defect. The higher percentages of normal dogs occur when one parent is normal and the higher percentages of more severely afflicted dogs occur when one parent is severly affected. The percentages are roughly similar for either sire or dam, indicating an absence of a maternal effect.

TABLE 17

Grade of Hip Dysplasia Affliction for Offspring according to Grade of Parent

Offspring		Grade of parent Normal	1	2	3–4
Sire: Number		*682*	*22*	*22*	*130*
Percentage:	Normal	59	64	32	22
	1	23	14	27	29
	2	9	22	23	19
	3–4	9	--	18	30
Dam: Number		*586*	*131*	*75*	*96*
Percentage:	Normal	60	57	28	30
	1	23	24	27	25
	2	10	8	12	16
	3–4	7	11	33	29

The third and most interesting piece of evidence is that given by selection programmes designed to reduce the incidence of the affliction. The simplest and most practical programme is not to breed from the most severely afflicted dogs and not from even mildly afflicted unless this cannot be avoided. The outcome of several programmes have been published, in some cases with credible results. In the German Shepherd Dogs of Sweden, the percentage of severely afflicted fell from over 30 per cent to under 20 per cent over a 5-year period. A similar programme in Switzerland reduced the percentage from about 45 per cent to just under 30 per cent over an 8-year period. In the German Democratic Republic, the percentage of afflicted animals was reduced from 44 per cent to 12 per cent over an 8-year period. All of these reductions were achieved by mass selection, which is relatively effective when the incidence of a trait is high, but less so when the incidence is low. That is, the best results occur in the early years; thereafter further progress is slow.

Since hip dysplasia is controlled by polygenes, straightforward or mass selection is the only method of tackling the problem of elimination of the defect. In any programme, it is unwise to breed from any animal which has been graded as severe. Those with moderate or mild grades should only be bred from if other attributes are outstanding. Testing for animals which are heterozygous for hip dysplasia polygenes is not feasible. However, a critical watch should be kept for any dog which is siring above average numbers of offspring with hip dysplasia. Such an animal should be immediately withdrawn from breeding (Henricson *et al.*, 1966; Hutt, 1967; Leighton *et al.*, 1977; Lust and Farrell, 1977; Bargai *et al.*, 1988). A number of serious attempts have been made to estimate the heritability – the proportion of the variation which can be attributed to heredity. These efforts have yielded rather variable results, ranging from around 20 per cent, through 30 to 45, and in one investigation as high as 60 per cent (summarized in detail by Willis (1989)).

Willis encapsulated these findings by saying that a reasonable assumption would be that between 25 and 40 per cent of the variation can be attributed to heredity. It is valuable to have some idea of the genetic determination since the success of any broadly based programme of selective breeding will depend on the heritability. If this is low (a value less than 10 per cent), the selection will have little effect. With a moderate value (say, 20 to 50 per cent), some progress will be evident. A high value (above 50 per cent) should produce worthwhile results. The estimates will display variation, due in part to the method of estimation – some are more accurate than others – and in part to the particular breed. The proportion of the variation which could be said to be genetic will differ from breed to breed, or even between strains within a breed. However, whatever the estimate may be, it is still wise to reject any animal with a high HD score or which is producing pups with high scores (Leighton *et al.*, 1977, Hedhammar *et al.*, 1979;

MacKenzie, 1985; Andersen *et al.*, 1988). An exhaustive review of the problem of hip dyslasia may be found in Willis (1989), who has critically assessed both genetic and non-genetic aspects. HD must surely qualify as the most intensively investigated of canine anomalies! As such, the problem is a fine example of the complex interaction between hereditary and environmental influences on the development of an anomaly.

Multiple Epiphyseal Dysplasia

This is an abnormal condition of epiphyseal calcification, that of the leg bones being particularly noticeable. The affliction is present at birth, causing difficulty in walking and a swaying gait. The joints may appear to be swollen. Surviving pups may exhibit stunted growth. The condition appears to be familial, but the precise genetics are unknown (Rasmussen, 1972).

Occlusion of the Teeth

The precise meeting of the teeth – bite – is an important aspect of the dog's physiognomy. The positioning and meeting of the teeth with mouth closed determine the bite. Five forms of bite may be recognized and these are the undershot, reverse scissor, level or pincer, scissor and overshot. The normal bite is the scissor, where the teeth of the upper jaw lie just in front of those of the lower. However, the level or pincer bite is common, where the teeth meet exactly, like the jaws of a pair of pincers. The reverse scissor is where the teeth of the upper jaw lie just behind those of the lower. Should there be a small gap between the two sets of teeth, the bite is said to be overshot when those of the lower jaw are behind the upper, or undershot when those of the lower jaw are in front of the upper. The bite of the Boxer and Bulldog are examples of the latter form of bite.

The occlusion is determined by the relative lengths of the upper and lower jaws. Any unsychronized development of the jaws will result in one of the above bites. Some of the variation will be non-genetic but a proportion will be genetic, produced by polygenes controlling the growth of the jaws. It is indeed very likely that some of the same polygenes which determine the many differences of skull shape for the various breeds are involved. Bite should be considered as part of the overall variation of head conformation.

The lower jaw (mandible) is subject to variation leading to malformation. Isolated cases could arise from accidents of development but, if a number of cases are observed in a stud, these may well have a genetic basis. Phillips (1945) reported a number of cases of overshot jaw in the Cocker Spaniel which he called "pig jaw." The condition occurred irregularly and he concluded that a recessive gene was involved with some degree of impenetrance. Unfortunately, the litters with afflicted pups were small and many of the pups died at an early age, which meant that the recessive inheritance could not be established conclusively. On the other hand, Gruneberg and Lea's (1940) analysis of a grossly shortened mandible in the Dachshund revealed an almost perfect assortment of a recessive gene *sm*. The anomaly interfered with normal mastication and there was some mortality.

Patella Luxation

This is dislocation of the kneecap and is an affliction of the forelegs. When one leg is affected, the dog walks or runs on three legs. If both legs are affected, the condition is serious since the animal is severely handicapped. The dislocation can be corrected, but the dislocation often recurs. The cause of the dislocation is due either to a too shallow groove in the femur in which the kneecap slides, or inadequate ridges on each side of the groove which help to retain the kneecap. In principle, the luxation can occur at any age, but it usually arises in dogs aged about 4–6 months. The degree of luxation may vary from a partial dislocation to one in which the joint becomes immovable.

Luxation of the patella could occur in any breed, but is found most often in the smaller breeds: the Miniature and Toy Poodle being particularly susceptible (Loeffler and Meyer, 1960; and Hodgman, 1963; Priester, 1972a). The inheritance of the defect is that of a threshold character, the pertinent polygenes controlling the shallowness of the groove in the femur, the development of the guiding bony ridges and the strength or attachment of the ligaments governing movement of the patella. All of these factors are determined polygenically and contribute towards the onset of the defect; not necessarily equally in all breeds.

Perthe's Disease

This affliction involves the head of the femur which becomes necrotic and partially disintegrates. There is some natural repair which may or may not be successful. It is a defect of young animals of the smaller breeds as a rule. The clinical symptoms are a sudden onset of lameness of the hind legs. There is some indication of pain being felt by the dog. Fortunately, surgical treatment is usually successful. The high incidence of the anomaly in certain strains is implicit of a genetic predisposition. Hickman and Spickett (1965) found evidence of this in the Miniature Poodle, Pidduck and Webbon (1978) in the Toy Poodle, Wallin (1986) in the West Highland White Terrier, and Robinson (unpublished) in the West Highland White Terrier and Yorkshire Terrier. The indications are that the defect is due to a recessive gene *pd*, possibly with variable penetrance, but it is not impossible for it to be inherited as a threshold character, a mode of inheritance which at times can mimic monogenic heredity.

Pituitary Dwarfism

Pups with this affliction are of normal size and appearance until about 1 or 2 months of age, whence a marked slackening of growth may be observed. The dwarfs are well proportioned. The coat retains much of its baby appearance, with some loss of hair, and is described as woolly-like. Growth soon ceases altogether, and death occurs at an early age. The growth retardation is due to under-functioning of the pituitary gland.

The dwarfism is produced by a recessive gene *dw*. Homozygous dwarfs are deficient in the growth hormone somatomedin, while the presumed heterozygote *Dwdw* have somatomedin levels intermediate between those for dwarfs and normal. The dwarfism was observed in the German Shepherd Dog in Denmark and Australia and in the Carelian Bear Dog of Finland (Willeberg *et al.*, Andresen and Willeberg, 1976a, b; Nicholas, 1978).

Polyostotic Fibrous Dysplasia

This condition has been reported in the Dobermann and Old English Sheepdog and is characterized by swellings just above the wrist of the forelegs. This is accompanied, or is followed, by lameness at about 5–7 months of age. The condition arises from cysts in the lower part of the radius and ulna. The bone material is replaced by fibrous tissue which contains pockets of degenerative matter and straw-coloured fluid. The condition may be successfully treated by surgery. No formal evidence was given for the assortment of a recessive gene. However, the incidence of the condition is familial and occurred in inbred animals. A genetic basis is suspected (Carrig and Seawright, 1969; Watson and Dixon,1977).

Pseudoachondroplastic Dysplasia

This achondroplasia is not apparent at birth, but becomes evident at about 2–3 weeks of age. The affected pup shows retarded growth and an inability to walk or stand. The long bones are short, thick and bent and have enlarged joints. The rib cage is flattened and leads to difficult breathing. The epiphyses are grossly abnormal and after growth is completed these ossify. General health is good except for arthritis. The genetic evidence is scanty, but monogenic recessive heredity is depicted (Riser *et al.*, 1980).

Short Spine

The spinal column is remarkably short, giving the animal a "hunched up" appearance. The shoulders are high and the back slopes sharply towards the tail. The trunk is crooked and the tail is short and screwed. The dog often sits on its rump with the front legs extended, rather in the manner of a baboon, a habit which gave rise to the popular name of "baboon dog". There was heightened mortality among these short-spined animals. The condition is stated to be produced by a recessive gene *sp* (Boom, 1965). Similar or identical genes have been reported from as far afield as Japan, South Africa and Sweden (Burns and Fraser, 1966; Hutt, 1979).

Skeleton Lethal

A syndrome of skeletal defects is produced by the dominant gene *Sl*. These include cleft palate, absence or extra digits to the feet, shortened hind legs, overshot jaw and wry-neck. Male pups with the affliction die within a few days of birth. However, female pups with the affliction survive and may be virtually of normal appearance or exhibit only mild skeletal defects. The litter data are very suggestive of sex-linked heredity (Sponenberg and Bowling, 1985). The anomaly occurred in the Australian Shepherd Dog.

Skull Fissures

The skull sutures of the newborn puppy can only be felt by careful examination. However, Pullig (1952) has reported instance of failure of the sutures to close (cranioschisis), appearing by manipulation as "soft spots", in the Cocker Spaniel. The soft crevices were about $\frac{1}{8}$ inch wide and $\frac{1}{4}$ inch long; instead of disappearing, these persisted or even became wider. None of the affected pups lived beyond 7 weeks. The data infer the assortment of a recessive gene *crs* for cranioschisis.

Slipped Disc

Slipped disc, or protrusion of the intervertebral disc, may occur in any breed, but the Dachshund is particularly prone to the anomaly. Slipped discs occur most frequently in mature dogs. Although no genetic studies have been made on the condition, the evidence for the liability by Dachshunds is soundly based (Gage, 1975; Priester, 1976). The analysis of the incidence in the breed by Ball *et al.* (1982) is indicative of polygenic threshold heredity for the disability.

Spondylosis Deformans

This condition refers to the outgrowth of bony protrusions on the edges of individual vertebrae. In time, these may fuse, causing an ankylosis or immovability of the vertebrae. The condition has been observed in many breeds, but a number are particularly prone to it. These are the Airedale Terrier, Boxer, Cocker Spaniel and German Shepherd Dog. The genetic aspect of the liability to the disease has been inferred from the differential frequencies between breeds (Morgan *et al.*, 1967).

Tail

The tails of dogs vary considerably in length, thickness and degree of taper. Many of the different types are illustrated by Whitney (1947). That tail form is inherited is shown by the many differences between breeds which are correctly held to be breed characteristics. Whitney's observations point towards polygenic inheritance for tail shape. Or, if major genes are involved, their expression is so variable that it is nigh impossible to follow through their inheritance.

Whitney noticed that tail length was inherited independently of leg length, so that some of his dogs had lengths of tail and leg which were rather incongruous. The drawback with Whitney's observations is that these were not carried through in a genetically systematic manner. If they had been, the results may not have been so apparently unfathomable.

The above remarks do not mean that major genes for certain tail conditions cannot be identified with favourable breeding data. For example, Pullig (1953, 1957) found that a tailless condition and a short tail with bends are inherited as if due to recessive genes *an* (anury) and *br* (brachyury), respectively. Also, Curtis *et al.* (1964) appears to have isolated a dominant gene *St* (stub) for a short tail. The complication here is that *St* has incomplete expression, so that some *Stst* animals have apparently normal tails. Some stub animals also had spina bifida.

The above results indicate that a tail anomaly cannot be simply classed as dominant or recessive without supportive breeding records. The tail is an easily damaged organ and an isolated instance of kinked, short, bobbed or even absence is usually due to accident, not heredity.

Thoracic Hemivertebrae

Signs of weakness of the rear legs and tail may be seen from about 6 weeks of age as a feature of the anomaly. Deterioration is rapid and within a few weeks the affected individual can scarcely walk. The muscles of the hindquarters display wasting. Radiographic examination revealed faulty development of the thoracic vertebrae which are clearly distorted. There is a narrowing of the vertebral canal and compression of the spinal cord. The frequency of occurrence of affected pups is in good agreement with assortment of a recessive gene *tr*. The anomaly was observed in the German Short-haired Pointer (Kramer *et al.*, 1982).

Toe Necrosis

Afflicted dogs of about 4 months of age began to lick their toes and eventually to bite at them. Claws were lost, followed by toes. There was a loss of sensitivity as shown by lack of reaction to cuts, pricks or heat. The animals failed to limp in spite of severely diseased toes. None lived beyond a year as a result of infection of the affected feet. A genetic basis for the affliction was proposed because of the inbred nature of the family of short-coated Setters in which it was observed (Sanda and Krizenecky, 1965). The data are certainly suggestive but inadequate to demonstrate conclusively that a recessive gene was involved. Similar cases of lack of sensitivity and self-mutilation of the toes have been described by Cummings *et al.* (1981*b*) for the English Pointer. Recessive inheritance is suspected and would be in keeping with the observed data but, again, these are insufficient to be conclusive.

Vertebrae Osteochondrosis

This is a defect of the spinal vertebrae. There is damage to the intervertebral discs and erosion of the cartilage end-plates of the vertebrae. This is followed by calcareous outgrowths on the ventral surface which fuse. The clinical signs are weakly developed spinal muscles, stilted gait, inability to fully extend the legs when running and to leap over fences easily cleared by a normal dog. Precise data on the mode of inheritance are lacking, but a genetic disposition is shown by the fact that all of the afflicted animals were sired by one stud. Pups sired by other studs of the colony in the same year were unaffected. Other afflicted animals have occurred previously in the colony of Foxhounds upon which the above observations were performed (Hime and Drake, 1965).

Neuromuscular Maladies

Defects of behaviour and muscles are usually traced back to various types of neural degeneration, either of the spinal cord or of the brain matter. The muscles are involved as a secondary factor, since these are controlled by the nerves. Any defect of the nervous system immediately affects behaviour and muscular control. For example, paralysis due to neural degeneration usually leads to lack of muscular tone and wasting away as a result of lack of use. A short review of a number of inherited behaviour pathologies is that of Griffiths (1986).

Ataxia

The ataxia manifests between the ages of 2 and 4 months of age when the afflicted pups begin to bark and howl unusually, rub their heads against objects and come to rest upon elevated places. They seem to experience difficulty in negotiating steps (down or up) and a bouncing locomotion becomes evident. After a short period, the symptoms stabilize and are not fatal. The ataxia is due to degenerative changes in the central nervous system. It is produced by a recessive gene *at* and was observed in the Fox Terrier (Bjorck *et al.*, 1962).

Ataxic demyelination

The condition develops very slowly, commencing at about 18 to 30 months of age (average about 26 months) as a general clumsiness and dragging of the feet. The rear limbs are usually the first to be affected but eventually the front limbs become the most severely afflicted. The ataxia becomes progressively worse, with swaying gait, goose stepping, stumbling and abnormal positioning of the feet, such as narrow or wide placing and crossing of the legs, and loss of stance during urination and defaecation. Some animals experience difficulty in rising and standing, and are reluctant to move, although their muscular tone seems to be good. The affected dogs remain bright and alert throughout, displaying no signs of distress (Wouda and Nes, 1986). The ataxia is inherited as a recessive and the gene involved is symbolized as *am*. The anomaly was observed in the Rottweiler.

Bjerkas' Leukodystrophy

This defect is yet another which manifests in ataxic-like behaviour of first the hindlegs and later all four legs. Eventually the limbs become paralysed. The affected pup also slowly loses its eyesight and staggers into obstacles. The early symptoms arise between 3 and 6 months of age. These symptoms were traceable to the formation of cavities in the white matter of the brain; the optic nerves and spinal cord were also involved. The condition is due to a recessive gene *ld* and observed in the Dalmatian (Bjerkas, 1970).

Cerebellar Ataxia

This ataxia made its appearance between the ages of 9 and 16 weeks. Initially affected pups develop a fine head tremor and stiffness of the hindlegs. This progresses until walking becomes difficult and falls are frequent. Eventually, movements of the head become exaggerated and the pup cannot stand without assistance. The condition is due to degenerative changes in the cerebellar cortex, with loss of Purkinje's cells. The ataxia was reported in the Kerry Blue Terrier and is due to a recessive gene *cb* (Lahunta and Averill, 1976).

Cerebellar Ataxia Hydrocephalus

With this disorder, abnormal symptoms develop between 4 and 6 weeks of age in the form of an unsteady gait, weakness of the hind limbs and visual problems. This latter involved bumping into obstacles, inability to locate objects and negotiating thresholds. Some pups showed a head tremor while others showed various compulsive behaviours. The cerebellar was degenerative, with numerous empty spaces in the tissues, together with an internal hydrocephalus. Pedigree data revealed that the lesions were produced by a recessive gene *cbh*. The defective pups were observed in a strain of Bull Terriers (Carmichael *et al.*, 1983).

Cerebellar Cortex Atrophy

Affected dogs behave normally until about 6 to 10 months of age when changes may be seen in the gait. The animal adopts an unsteady, prancing and over-reaching walk which slowly increases in severity. A lack of proper co-ordination of the limbs becomes more evident. The rear legs are held wide apart. The cortex of the cerebellum was found to have undergone degenerative changes, with loss of Purkinje cells. The anomaly was inherited as a recessive (gene symbol *cca*) and was observed in the Gordon Setter (Lahunter *et al.*, 1980; Cork *et al.*, 1981; Steinberg *et al.*, 1981; Troncoso *et al.*, 1985).

Cerebellar Degeneration

The first signs of disorder appear at about 4–8 weeks of age and consist of lack of co-ordination of the hindlegs. This increases in severity and the pup adopts a stance with the legs apart. Some animals develop a tremor of the head, which becomes more pronounced when walking or running. The sense of balance is disturbed and falls are frequent, with the legs fully extended. Severely affected individuals spent most of their time resting, but were able to climb to their feet unaided. Autopsy revealed progressive degeneration of the lobes of the cerebellum and loss of granula and Purkinje cells. The condition may be abscribed to a recessive gene *cbd* although the data are not as extensive as might be (Hartley *et al.*, 1978).

Cerebellar Neuroaxonal Dystrophy

The malady was seen in Collie Sheepdogs of Australia and New Zealand. Affected individuals showed abnormal symptoms at about 2 to 4 months of age which increased in severity. These included a wide-based stance, accompanied by a high-stepping walk. All had a persistent tremor and displayed difficulties in maintaining balance. Numerous spheroid bodies were detected in the central cerebellar and adjacent peduncular white matter of the brain. Circumstantial evidence suggests the condition is inherited as a recessive (Clark *et al.*, 1982).

Epilepsy

It is probable that many, if not most, cases of epilepsy are due to heredity. The problem of determining the mode of inheritance, however, has proved to be complex. An initial suggestion that epileptic fits are determined by a recessive seem to be false. Three careful studies by Falco *et al.* (1974) for the German Shepherd Dog, Wallace (1975) for the Keeshound, and Cunningham and Farnbach (1988) for the Beagle have failed to uncover any simple mode of inheritance. On the other hand, Edmonds *et al.* (1979) for the Beagle and Willis (1989) for the Welsh Springer Spaniel concluded that a single recessive gene is involved, with some degree of impenetrance. In all breeds with an incidence of epilepsy, males are more susceptible than females. A major gene could be involved but it may be assorting against a background of threshold heredity where the threshold for seizure varies for the individual. Breed, age, sex and inciting stimuli have emerged as environmental factors which mediate the condition.

Fucosidosis

The anomalous condition results from a deficiency of an enzyme which is necessary for efficient functioning of the nerves. Individuals with the deficiency are normal until approximately 6 to 12 months when they commence to show apprehensive behaviour. This deteriorates to a high-stepping gait and unusual positioning of the feet, leading to severe incoordination of movement. There is a diminution of vision and an intermittent nystagmus. A few dogs have difficulty in eating, regurgitating food and losing weight. The females are fertile but the males are sterile due to reduced numbers and grossly abnormal sperm in the ejaculate.

Necropsy findings reveal gross lesions of the nervous system. Afflicted dogs are in reasonably good health until the advanced stages of about 24 to 30 months. The malady is inherited as a recessive in the English Springer Spaniel (Healy *et al.*, 1984; Barker *et al.*, 1988; Taylor and Farrow, 1988; Taylor *et al.*, 1989) and is given the symbol *fu*. A proportion of heterozygotes can be detected biochemically which would be valuable information should an eradication programme for the gene ever be necessary.

Giant axonal neuropathy

The first sign of the anomaly is a dragging of the toes of the hind legs at about 14 to 16 months of age. The behaviour degenerates into a

progressive ataxia, with uncoordinate movements and finally paralysis. Muscle tone is depressed, with wasting below the stifles. The front legs escape much or all of the deterioration. The barking reflex is diminished. From approximately 18 months, there is regurgitation of undigested food as a consequence of a swollen oesophagus. Faecal incontenence also occurs. The coat is tightly curled in contrast to the normal smooth condition. The characteristic of the malady is the giant or swollen appearance of the axons of the nerve fibres. The anomaly occurred in the German Shepherd Dog as a recessive trait (Duncan and Griffiths, 1981) and may be symbolized
by *gan*.

Globoid-cell Leukodystrophy

With this defect, the first signs are observed within a few weeks to months of age, as a lack of co-ordination of movement. The affected animal may stumble, hold its hindlegs apart as if to maintain balance, or collapse because of loss of control. A tremor is usually present, slight at first, but becoming more pronounced. Autopsy revealed large numbers of globoid cells throughout the white matter of the brain; eventually most of the white matter is destroyed. A recessive gene *gc* is responsible for the defect. The same or similar defect has been identified in a number of breeds (Hirth and Nielsen, 1967).

GM1 Gangliosidosis

Dogs with this metabolic disease become ataxic at about 2 to 4 months of age, displaying uncoordinated and exaggerated movements of the legs, hyperactivity and excitability. The condition deteriorates steadily until the animal cannot stand upright. A head tremor may be evident. Extensive lesions can be seen in the brain tissues. The disease was observed in a colony of mongrel Beagle type dogs by Read *et al.* (1976), who found that the malady was inherited in a recessive manner.

A probably identical disease is described by Alroy *et al.* (1985) in the English Springer Spaniel. These dogs were undersized, showed progressive abnormal behaviour, brain degeneration, head bossing and deformities of the long bones. Here, too, the condition showed recessive heredity and the causative gene may been symbolized by *ga1*.

GM2 Gangliosidosis

The onset of this defect is at about 6 months of age and the anomaly manifests as increasing nervous behaviour and impaired response to training. At about 9 to 12 months, walking and running become clumsy or ataxic, accompanied by occasional fits. Partial blindness and deafness also develop and affected dogs tend to lose sensory contact with their surroundings. Death usually occurs at about 2 years of age.

Pedigree analysis indicates a recessive mode of inheritance and, because all the affected pups were male, it was speculated that the *ga2* gene is sex-linked. However, the number was too few for the speculation to be accepted uncritically and the heredity is now considered to be ordinary autosomal. The defect was observed in the German Short-haired Pointer (Karbe and Schieffer, 1967). The abnormal behaviour results from defective enzyme activity which permits lipid material to accumulate in the brain cells.

Hypertrophic neuropathy

Pups with this anomaly show a generalized muscular weakness, depresed reflex action and loss of muscular tone. The first signs are apparent between 7 and 12 weeks of age and become progressively more severe. The expression varies from a shuffling gait to complete collapse. The malady was detected in a kennel of Tibetan Mastiffs and is due to a recessive gene *hn* (Cummings, 1981; Sponenberg and Lahunter, 1981; Cooper *et al.*, 1984).

Muscle Fibre Deficiency

The clinical signs of the deficiency appear between 3–5 months of age. The initial signs are an inability of the affected dog to hold up its head. This is followed by general muscular weakness of the whole body, and pria pism in the male. Exercise, cold weather and excitement would aggravate the condition. With continued exercise, the limbs would become extended, the gait consisting of short mincing steps. The hindlegs come together and the animal adopts a hopping gait. Frequently, the front legs would collapse, causing the dog to slither forward. After a few minutes of rest, the condition would abate, only to reappear upon the resumption of exercise. There is a stunting of growth and a deficiency of muscular mass. Detailed studies revealed the latter to be due to a deficiency of type II muscle fibres. The condition was found in a strain of Labrador Retrievers in America by Kramer *el al.* (1976, 1981) and has been assigned the symbol *mfd*. A clinically similar disease of muscular weakness has been described for Labrador Retrievers in England. However, it is not clear whether or not this is identical to the American anomaly because there are pathological differences (McKerrell *et al.*, 1984).

Myasthenia Gravis

This is a form of muscular weakness which is brought on by prolonged exercise but which may be alleviated by a period of rest. The young pup is least affected although it may appear less playful than usual. On the other hand, adult animals quickly show signs of extreme fatigue; their walking or running falters, followed by a collapse. After a rest, they can resume normal activity until exhaustion sets in once again. There is no sign of muscular wasting. The anomaly has been reported for three breeds: Gammel Dansk Honshund (Flagstad, 1982), Smooth Fox Terrier (Miller *et al.*, 1984) and Jack Russell Terrier (Wallace and Palmer, 1984). In each of these breeds, the condition is inherited as a recessive. In the Smooth Fox Terrier the gene has been symbolized by *mg*. There is no means of ascertaining if the same gene is responsible for the myasthenia in each breed but this may be provisionally assumed.

Myelopathic Paralysis

This paralysis develops between the ages of 3–8 months. It is characterized by a weakness beginning with the hindlegs and then involving the front legs. The weakness progresses to general and increasing unsteadiness until the limbs become totally paralysed. The condition is due to necrosis of the myelin sheath of the primary nerve fibres of the thoracic region of the spine. It has been observed on at least two occasions in the Afghan Hound and is produced by a recessive gene *mp* (Averill and Bronson, 1977).

Myotonia

This is a defect characterized by over-contraction of the muscles. The early symptoms are stiffness of the legs, appearing at about 3 months of age. The forelegs are held apart and the rear legs are bunched together. As the animal moves forward, with the forelegs extended, the body sways from side to side. After a few paces, the muscles relax and the dog moves normally; lifting the animal would cause an immediate spasm. The symptoms persist but become less severe after a year or so. The defect occurred in the Chow Chow, Irish Terrier and a few other breeds. It has been speculated that the anomaly is due to a recessive gene but confirmatory data are desirable (Wentinck, 1974; Farrow and Malik, 1981; Jones, 1984).

Narcocataplexy

Dogs with this disorder appear normal except for attacks of either spontaneous or induced paralysis of the limbs. The induced attacks may be elicited by excitement, such as playful behaviour with either other dogs or human companions, the offering of tit-bits or sexual activity. The onset of episodes varied widely from 1 to 4 months of age, with exceptional individuals not experiencing the first attack for over a year or two.

Four independent instances of the anomaly were investigated by Baker *et al.* (1982) and two, those in the Dobermann Pinscher and in the Labrador

Retriever, proved to be genetic. In each breed, the cataplexy was found by breeding experiments to be inherited as a recessive. Furthermore, crosses between susceptible animals from each breed produced susceptible progeny. This implies that the same gene (*nc*) is involved in each instance. The average age of onset of the cataplexy differed between the breeds, being 8.6 weeks for the Dobermann but 14.4 weeks for the Labrador. The average for the cross-breeds was 9.2 weeks, roughly midway between the averages for each of the pure breeds. Clearly, there are genes which can modify the manifestation of the disorder.

The other two instances were investigated genetically but yielded no evidence of a genetic basis to the cataplexy: these were affected individuals of the Beagle and Miniature Poodle breeds. Although a genetic involvement cannot be fully excluded, their descendants appeared to be completely normal. For many anomalies, it is possible for genetic and non-genetic cases to closely resemble each other phenotypically (as a geneticist would say) or clinically (as a veterinarian would say). The present investigation is a notable example where it has been possible to distinguish between the two.

Neuroxonal Dystrophy

This is a neurological disorder which develops progressively from about one year of age. The gait is mainly affected in the form of a swinging prancing walk, most evident in 3-4-year-old animals. The dog commonly stands awkwardly and shifts its weight from one foot to another. There may be an unwillingness to jump over low obstacles. Making turns seems to create difficulties, as well as ascending and descending stairs. In fact, by four years, the ability to negotiate stairs seems to have been lost. Pathologically, the disorder appears to be due to enlargement of the distal axons which are filled with inclusions (Cork *et al.*, 1983). The anomaly was seen in the Rottweiler and could be inherited as a recessive.

Neurogenic Muscular Atrophy

The initial signs of the neurological malady begin as a tremble in the upper rear limbs at about 18–23 weeks of age. Affected dogs show clear symptoms of progressive muscular weakness. They may stumble and make frequent stops to rest when out walking. Finally, as the disease progresses, they are unable to stand, even with assistance. The muscles commence to waste away and this is most evident for the extremities. Although the dog is fully recumbent, it remains alert and can move the head and wag the tail. There is no loss of control of micturition or defaecation. However, there is a cessation of growth in the final stages. The discorder has its origins in a degeneration of the lower motor neurons, certain of which showed granular inclusions (Inada 1978, 1986). The anomaly was observed in the English Pointer and is produced by a recessive gene *nma*.

Neuronal Abiotrophy

This is a devastating loss of movement of the leg muscles. The symptoms first appear at about 5–7 weeks of age, when weakness of either the front or rear legs may be noticed. Within a week, all four limbs may be affected and soon the pup cannot stand. It will lay on its chest, with the legs either bent under or extended. The leg muscles quickly waste away, although a few animals can crawl feebly. The trunk muscles are only slightly affected so that the pup can move its head and even wag its tail. The defect does not worsen further, but is for most purposes lethal. The cause is degenerative changes in the central nervous system. It is due to a recessive gene *na* with regular manifestation and discovered in the Swedish Lapland Dog (Sandefeldt *et al.*, 1973).

Neuronal Ceroid-lipofuscinosis

Affected dogs show normal behaviour until about 12 months of age, when the animal increasingly fails to respond to commands, movements show signs of unco-ordination and it becomes partially blind and deaf. In the final stages, muscular spasms appear and the dog may have a problem in locating its feed. The animal becomes dehydrated and emaciated, and it is rare for any to live beyond 2 years. Autopsy findings indicate extensive deposition of lipid granules in the nerve cells of the central nervous system.

The defect is reported for the English Setter and is produced by a recessive gene *li*. A deficiency in phenylenediamine mediated peroxidase activity has been found between normal animals and those heterozygous for *li*. This means that the normal appearing heterozygotes can be identified should this be necessary (Koppang, 1970; Patel *et al.*, 1974).

A similar, if not an identical, condition is described by Appleby *et al.* (1982) for the Saluki. The breeding data are meagre but suggestive of monogenic heredity.

Progressive Axonopathy

In this ataxia, the first symptoms appear about 6 months of age, exceptionally later, as a swaying, unco-ordinated gait of the rear legs, which often cross when walking. There are signs of muscular weakness, but little or no wasting. The anomaly is due to pathological changes in the central nervous system, some peripheral denervation, but especially axonal degeneration of motor nerve fibres (Griffiths *et al.*, 1980). The anomaly appears to be confined to the Boxer and is due to a recessive gene *pa* (B. M. Cattanach, 1981, personal communication).

Quadriplegia

This is the condition commonly called "swimmers", because affected pups are unable to stand but lay on their stomachs and propel themselves by lateral or swimming motions. The condition is congenital and can be recognized as early as 3 days of age. There may be head tremor, and movements are often jerky and uncontrolled, accompanied by occasional fits. The neck muscles appear to be weak, but later the pup can eat if the muzzle is placed in the food. However, eating behaviour is unusual; often the jaws will clamp tightly on a mouthful of food and it is some time before chewing commences. The pups are totally or partially blind and a nystagmus is present. Hearing is normal. Lesions are observed in the cerebelum, with degeneration of the Purkinje cells. The condition is due to a recessive gene *qu* (Palmer *et al.*, 1973).

The above observations were carried out upon a strain of Irish Setters. However, Hutt (1979) has noted that the term "swimmers" has been used to denote any condition where the pup lays prone upon its belly because of an inability to stand or walk. He notes that not all swimmers are blind, and that some recover to become normal dogs, whereas the quadriplegias did not. These comments emphasize the point that similar conditions in the same or different breeds (especially the latter) may not be due to the same gene. Small differences of expression are often an indication of this. However, the expression of the same gene could be different in different breeds. This means that each case must be examined individually and assessed on its own merits.

Scottie Cramp

This is an extreme rigidity of the muscles brought on by excitement or strenuous exertion. Dogs at rest or during light exercise do not display spasms. During an attack, the afflicted animal arches its back, draws up its hindlegs while extending the front legs. A few animals are able to walk by a hopping gait, but the majority cannot move. These latter may stand as if in a trance or fall over, laying motionless on one side. Consciousness is not lost. There is a return to normality if the dog is allowed to recover by itself. The condition may manifest as early as 6 weeks of age and is usually fully developed by 1 year. The condition is produced by a recessive gene *sc* and the designation "Scottie cramp" is due to the fact that it was first studied in the Scottish Terrier (Meyers *et al.*, 1970).

Sensory Neuropathy

An anomaly has been observed in the long-haired Dachshund where the legs assume unusual positions, accompanied by minor staggering, an apparent reduction or lack of sensation of pain, and urinary and faecal incontinence. The behaviour appears to be due to a severe depletion of myelinated fibres in the sensory nerves. The affected animals were related and a genetic aetiology is probable (Duncan and Griffiths, 1982).

Sex-linked Myopathia

This is a degenerative disease of the muscles. The first symptoms appear at about 8 weeks of age in the form of an enlarged tongue, accompanied by difficulty in swallowing. After several weeks the enlargement of the tongue is very conspicuous and the mouth can only be partially opened. The affected pups can only walk stiffly, their backs become arched and they are unable to rear onto their hindlegs or jump over obstacles. Autopsy revealed that the muscle tissue was pale, with yellowish-white streaks. The condition is due to a sex-linked recessive gene *my*. Heterozygous females are of normal appearance and affected pups are males bred from these females. The defect has been observed in the Irish Terrier (Wentinck *et al.*, 1974).

Shaking

The feature of the malady is a persistent tremor which involves the whole body, head and limbs, manifesting at about 10–12 days of age. When the pup is sleeping or lying quietly, the tremor is barely perceptible or absent but it reappears on arousal. Their movements are unco-ordinated and they are unable to stand. When attempting to move, the pup slithers sideways or backwards. The entire central nervous system was found to be grossly abnormal and particularly deficient in myelin sheathing of the nerves. Affected individuals were only about half the size of normal litter mates and their deterioration was such that they had to be humanely killed. The condition is inherited either as an autosomal recessive or as a sex-linked trait (Griffiths *et al.*, 1981). The latter is probable since only male pups were affected in three litters sired by two different males. Unfortunately, the numbers are too few to be decisive. The gene is symbolized by *sh* and was found in the Springer Spaniel.

Spinal Dysraphism

A locomotive defect which occurs between 4–6 weeks of age. The symptoms are variable and include a characteristic hopping hindleg gait when walking and a crouching posture as if frightened, with the hindlegs spread out and extended backwards when standing. The symptoms tend to disappear as the pup matures. The condition is due to cavities and other lesions in the spinal cord. In early studies it was thought that the condition was due to a recessive gene with normal overlaps, but later work is suggestive of a dominant gene with normal overlaps, (Draper *et al.*, 1976). There is little to chose between these genetic alternatives when one realizes that normal overlaps mean dogs which carry the genetic liability for the defect but do not exhibit symptoms themselves. Further investigation is required to clarify the situation.

Spinal Muscular Atrophy

This is a muscular weakness defect which develops at about 2–8 months of age. The affected dog is disinclined to exercise and soon shows signs of marked exhaustion. There is often laboured breathing and excessive trembling. Later the gait becomes abnormal, the dog adopting a crouched waddling movement of the pelvis. Eventually, the animal cannot stand. There is loss of weight and a wasting away of the muscles in the pelvic region. Histological findings indicate a progressive deterioration of the innervation of the muscles of the lower spine. The defect was observed in the Brittany Spaniel and is acribed to the action of a recessive gene *sma* (Cork *et al.*, 1979; Lorenz *et al.*, 1979).

Stockard's Paralysis

This paralysis affects the hindlegs. The condition results from a loss of function of the neurons arising in the lumbar region of the spinal cord. The defect manifests about the 11th to 14th day when the pup may collapse or have difficulty in walking. The hindlegs cannot be extended properly and a rambling gait is affected. Defective individuals have been observed in the Great Dane and in the St. Bernard. The paralysis has a genetic basis, but appears to depend upon more than one gene for its expression (Stockard, 1936). It is probably a case of threshold heredity.

Blood Disorders and Cardiovascular Defects

Disorders of the blood may be grouped into two main categories: anaemias, of which there are many forms, and haemophilia, which is a disruption of the normal clotting process. Again there are a number of different forms, all resulting in excessive, and frequently fatal, bleeding. The cardiovascular defects are anomalies of heart structure or associated major blood vessels. These are not necessarily severe enough to cause death by heart failure, but they can impair health.

Congenital Lymphoedema

Affected individuals have swollen limbs, the surface of which remains depressed upon the application of pressure. The swelling is not apparently painful. The extent of the oedema varied from the hindlimbs only, through all four limbs to a condition in which the body and tail were involved. The oedema is present at birth and accompanied by high mortality among those most severely affected. Thereafter, the expression of the oedema declined, with the least affected becoming of normal appearance. The accumulation of fluid is attributed to a defective lymph drainage system. The condition is due to a dominant gene *Ly* (Patterson *et al.*, 1967).

Cyclic Neutropenia

This gene has been previously discussed as CN dilution because of its effect on coat colour. Why the gene should lighten the coat colour is unknown. However, the *cn* gene is potentially lethal, for it induces cyclic fluctuation in the numbers of neutrophils at 10–12 day intervals. The neutrophils are a blood cell concerned with the body's defence against invading pathogenic bacteria. Thus, when the numbers of neutrophils are low, so is the resistance to infection.

The clinical symptoms are fever, loss of appetite, inflammation of the gums and joints, the latter leading to lameness, listlessness and various infections. Diagnosis may be based upon coat colour and analysis of numbers of different leucocytes. CN pups usually die within a few weeks of birth and nearly all are dead by 6 months. Life may be prolonged by treatment with antibiotics during the periods of neutropenia. These animals are undersized and frequently sterile (Ford, 1967; Cheville, 1968; Lund *et al.*, 1970; Campbell, 1985). The disease is also known as the grey Collie syndrome. Transplants of bone marrow cells greatly extended the life of affected animals and even caused the coat to assume its normal colour (Yang, 1978; Matsas and Yang, 1980).

Elliptocytosis

The red blood cells (erythrocytes) of this disease are of smaller volume and most are more oval than is normal. Their lifespan is significantly shorter than usual, an aspect which is compensated by an increase in their number. They also show greater fragility under stress. At present, the anomaly has not been studied sufficiently to discover if it results in an anaemia or manifests other untoward clinical symptoms. The changes in the corpuscles are produced by a deficiency in one of the membrane proteins. Technically, the protein is represented by band 4.1 in the appropriate electrophoretogram. The band is absent in the deficient individuals and appears as a trace in the heterozygotes. The anomaly is inherited as an incomplete recessive, since the presumed heterozgotes have a milder form of the elliptocytosis (Smith *et al.*, 1983), and may be symbolized by *ell*.

Granulocytopathy

An impairment of bactericidal function of the granular leucocytes has been investigated by Renshaw and Davis (1979) in the Irish Wolfhound. The clinical manifestation of the dysfunction is increased susceptibility to pyrogenic (fever-inducing) infection and a markedly reduced lifespan. The impairment was demonstrated by *in vitro* leucocyte assay of killing of *Escherichia coli*. Breeding experiments indicated that the dysfunction was inherited as a recessive, with gene symbol *gr*.

Haemolytic Anaemia

This anaemia was first found in the Basenji. The anaemia may be easily overlooked until it is too late for treatment. The anaemia is unremitting and progressive. The first signs appear at about 1 year and death follows within a year or two. Although appearing outwardly normal, the affected dog shows decreased activity and may resist being taken for exercise. The mucous membranes are paler than normal and the spleen becomes enlarged. However, the anaemia can be diagnosed by precise examination of the red blood cells. In fact, in this manner it is possible to detect the onset of the anaemia as early as 8 weeks. At this stage the anaemia behaves as a recessive to normal.

Further investigations revealed that the red blood cells were deficient in the essential enzyme pyruvate kinase (PK). The quantity of PK is governed by a gene *pk*. Assays of PK activity of normal dogs showed that these fall into two groups, differing average activity, one group showing greater activity than the other. These are the homozygote (*PkPk*) and heterozygote (*Pkpk*), respectively. Unfortunately, the range of variations overlap, so that a few animals cannot be reliably identified as either *PkPk* or *Pkpk*. However, significant differences were found for adenosine triphosphate content of stored red blood cells and for phosphoenolpyruvate assay between the two genotypes. The latter difference is evident for adults but not for puppies. The fact that heterozygotes *Pkpk* can be identified by special tests is an advantage in eliminating the defect (Brown and Teng, 1975; Andersen, 1977). The PK deficiency has also been reported for the Beagle (Prasse *et al.*, 1975).

A similar haemolysis has been discovered in English Springer Spaniels (Giger *et al.*, 1986; Giger and Harvey, 1987). The symptoms are somewhat different from the above form but show the typical lethargy and muscular weakness after exercise. The urine may be discoloured. The malady is less severe in that life expectancy is not markedly reduced. In the Spaniel dogs, the haemolysis is due to a deficiency of the enzyme phosphofructokinase (PFK). The deficiency is inherited as a recessive as regards clinical expression but heterozygotes can be detected biochemically. The gene may be symbolized by *pkf*.

A haemolytic anaemia has been described by Randolph *et al.* (1986) which resembles in general terms the disease induced by the PK deficiency but is apparently quite different. In spite of an extensive analysis, the cause could not be ascertained. The pedigree chart given by the authors indicated recessive heredity of the anaemia.

Haemophilia

Clotting of the blood is one of life's most important defences and is a surprisingly complex process, consisting of at least thirteen stages ("factors") which must be completed before normal clotting occurs. An interruption of any of these stages can result in mild to severe haemophilia. Actually, the process is even more complicated than the above statement would suggest. However, few breeders are probably interested in the physiological mechanism of blood clotting nor in the technical details of how the various forms are diagnosed. These aspects have been thoroughly treated by Hall (1972). Valuable summaries of the literature are provided by Dodds (1974, 1988), Spurling (1980), Grun (1985), Parry (1985), Johnson *et al.* (1988) and Fogh and Fogh (1988).

The extent of the haemophilia can vary from individual to individual as a result of chance factors, the innate efficiency of the individual's clotting mechanism and the nature of the disability. Different genes can produce characteristic maladies, but, since the clinical symptoms are restricted to failure to check indiscriminate bleeding, it is not surprising that different haemophiliac genes can induce similar symptoms. The different disorders can be differentiated, however, by the appropriate laboratory techniques.

The severity of the haemorrhagic symptoms have been defined as follows:

1. *Severe.* Prolonged bleeding after injury or surgery; recurrent spontaneous bleeding in muscle tissue and joints to such an extent to endanger life.
2. *Moderately severe.* Copious bleeding from injury or surgery, some bleeding from minor wounds; occasional spontaneous internal haemorrhaging, but without sufficient loss of blood to endanger life.
3. *Mild.* Bleeding only after major injury or surgery; occasional slight bleeding from minor injury.

The most important haemorrhagic diseases are summarized in Table 18. Haemophilia A is the most common form and at one time or another has been reported in most of the popular breeds. The causative gene has a farily high rate of mutation. It is interesting that both haemophilia A and B are carried by the *X* chromosome and are sex-linked. It is the males which

TABLE 18

The Major Haemophiliac Defects

Defect	Bleeding symptoms	Heredity
Haemophilia A (Factor VIII deficiency)	Moderate to severe	Sex-linked; milder effects in heterozygous females
Haemophilia B (Factor IX deficiency)	Severe	Sex-linked; milder effects in heterozygous females
Hypofibrinogenaemia (Factor I deficiency)	Severe	Dominant
Hypoproconvertinaemia (Factor VII deficiency)	Mild to moderate	Recessive; heterozygotes can be detected by laboratory tests
Hypoprothrombinaemia	Severe	Recessive
Plasma thromboplastin antecedent (Factor XI deficiency)	Moderate to severe	Dominant; effects much more severe in homozygotes
Stewart-Prower (Factor X deficiency)	Moderate to severe	Dominant; effects much more severe in homozygotes
Thrombasthenia	Moderate to severe	Dominant
Von Willebrand's disease	Mild to moderate	Dominant

are usually afflicted, produced by heterozygous or "carrier" females. Dogs with severe forms of haemophilia can usually be detected by haemotomas (soft swelling under the skin, full of blood), periodic lameness (due to bleeding in the joints), bleeding from the mouth, nose and anus, and, of course, profuse bleeding from any wound, out of proportion to the extent of the injury. Haemophilia with mild effects may pass unnoticed until such time as stress, such as injury, which calls for surgery or parturition. If haemophilia is suspected, a veterinarian should be consulted immediately. Only he can determine if the disease is haemophilia and set in motion the elaborate tests necessary to diagnose which sort it may be (if the latter is thought to be worthwhile).

The thrombasthenia of the table refers to the platelet defect studied by W. J. Dodds in the Otter Hound, but it seems that a similar yet different thrombopathy occurs in the Basset Hound (Johnstone and Lotz, 1979). In both cases, the defect is inherited as a dominant, in that the heterozygote can be detected by laboratory tests, but the clinical anaemia behaves as a recessive, in that only the homozygote displays classic symptoms.

Dogs with the Ehlers-Danlos syndrome not only have excessive elasticity of the skin but lacerate easily. There is poor wound healing, accompanied by bleeding. This is not strictly a haemorrhagic disease, although it could be taken as such because of the bleeding tendency.

Pelger–Huet Anomaly

Dogs with this anomaly have round, oval or bean-shaped nuclei in the neutrophils (one of the white blood cells) instead of the normal segmented nuclei. The condition appears to be harmless, but it is possible that resistance to disease is impaired. This could reduce survival of young pups and longevity of the adult. The anomaly is due to a dominant gene *Pg* and the above description is for the heterozygote. The homozygous Pelger is unknown, but in the rabbit the homozygote is a perinatal lethal. This does not augur well for the genetic situation in the dog (Bowles *et al.*, 1979).

Systemic Histiocytosis

The disease results from massive and abnormal infiltration of histiocytes into many tissues of the body. The most obvious clinical symptoms of the process are to be seen on the skin. Numerous fibrous nodules form all over the body but are most concentrated on the scrotum and about the nose and eyes. There is a loss of appetite, decline in weight, laboured breathing, and conjunctivitis. There is clear indication of the malady being familial but the precise mode of inheritance could not be determined (Moore, 1984).

Cardiovascular Defects

The last two decades have seen the publication of some significant findings on the nature and incidence of congenital heart defects. These are due in large measure to the studies of D. F. Patterson and his associates. It has been found that the majority of heart complaints can be accounted for by seven specific defects. These are patent arteriosus stenosis, pulmonic stenosis, aortic stenosis, ventricular septal defect, atrial septal defect, persistent right aortic arch and tetralogy of Fallot. Pure-bred dogs show a much greater predisposition towards heart defects than cross-breds or mongrels. Furthermore, the individual defects listed above occur more frequently in certain breeds than others; for some defects, merely in a few breeds. This sort of evidence indicates a genetic background to the incidence. Although many popular breeds are concerned, not too much should be made of this, *qua* breeds, because any breed may develop a liability for a certain heart defect.

Confirmation that the defects are inherited was shown by breeding experiments. The incidence rose sharply when dogs with specific defects were mated together. The inheritance was not monogenic, but of the polygenic threshold type. With some defects, the situation was a simple normal: defective dichotony, but in others it was possible to grade the severity of the defect according to the polygenes contributing to it. Thus, some defects had two or three developmental thresholds for increasing severity. These conclusions are not entirely academic, despite the fact that dogs with mild heart defects may show little or no ill-effects. They would, however, have a propensity to transmit the defect to their offspring, possibly in a more severe form.

The usual signs of a heart defect are a lack of stamina, breathlessness, weakness of the limbs, loss of weight, limb oedema, cyanosis and loss of consciousness. Not all of these symptoms need be expressed, depending upon the defect and the severity. Dogs with mild defects may appear normal unless given strenuous exercise. On the other hand, severe conditions can cause the pup to die within a few weeks and shortens the life expectancy of adults. Most of the conditions can be detected by listening for abnormal sounds emitted by the heart. Several of the conditions produce characteristic sounds and these can be interpreted by a specialist. Two comprehensive reviews are by Patterson (1974, 1976).

Miscellaneous Anomalies

These are a mixed group of abnormal conditions which cannot be easily placed in the foregoing categories. In general, they are anomalies of the soft tissues (other than muscles), internal organs and of the dog's physiology.

Autoimmune Diseases

The incidence of these diseases is probably low in the general dog population, and frequently their true nature may not be recognized. The body's defence mechanism against infectious disease is turned upon itself and certain organs are infiltrated and attacked by lymphocytes. The result is destruction of tissues, impairment of function and chronic inflammation.

In a large colony of Beagles, recurring cases of two forms of these diseases, viz. lymphocytic thyroiditis and lymphocytic orchitis, have been observed, the thyroid gland and testes being attacked, respectively. It was found that the cases occurred in certain blood lines which could be traced back to common ancestors. In view of this finding, it was concluded that both of these diseases were determined genetically, although the mode of inheritance could not be established (Fritz, *et al.*, 1976).

Conoaway *et al.* (1985) has reported spontaneously arising thyroiditis in a colony of Borzois. The anomaly was followed over three generations and is described in some detail. The thyroids showed inflammatory and degenerative lymphocytic changes which lead to fibrosis of the gland. The disorder is stated to be inherited as an autosomal recessive trait but no breeding data are presented.

Copper Toxicosis

Copper poisoning is an insidious disease in that the affected individual often does not display any symptoms until the malady is well advanced. High levels of copper accumulate in the liver due to a metabolic deficiency, and death eventually ensues from extensive cirrhosis. Typical signs are a general appearance of being "out of sorts", listlessness, a loss of appetite and poor condition, frequently accompanied by inhanced thirst and increased urination. Veterinary attention can prolong life but the outlook is poor.

The condition has been shown to be inherited as a recessive trait in the Bedlington Terrier (Johnson, *et al.*, 1980) and in the West Highland White Terrier (Thornburg, *et al.*, 1986). The gene has been symbolized by *ct* but it is not certain that the same gene is involved in each breed. An identical or similar disease has been reported for the Doberman Pinscher (Johnson *et al.*, 1982) but it is uncertain if the condition is inherited. Johnson *et al.* (1984) have perfected a technique for the detection of affected animals long before clinical symptoms are evident. This will enable these animals to be withdrawn from breeding before they have inadvertently passed on the *ct* gene.

Cryptorchidism

Cryptorchidism is the condition where either one (unilateral) or both (bilateral) testes fail to descend into the scrotum. The bilateral form is sterile, but the unilateral is usually fertile. The condition is sex-limited because obviously females cannot be afflicted. The defect is, or should be regarded, as a serious fault of the intact dog. A genetic liability to the defect has been suspected for a long time and some reports have even stated that it is inherited as a recessive (cf. Burns and Fraser, 1966). However, the data are not conclusive on the mode of inheritance, merely indicating that the condition can be familial. Although this is unsatisfactory, it is a little better than having no information at all. Treating the two forms of cryptorchidism as a single defect, it is probable that it behaves as a threshold character. That is, the condition is due to the action of several polygenes, none of which are capable of producing cryptorchidism by themselves, but acting in unison are capable of so doing. These polygenes are present in breeds at a low frequency, but may be concentrated by chance in certain strains, families or even individual animals. Then the incidence of cryptorchidism increases quite sharply and people begin to wonder why.

The sterility of the bilateral cryptorchid is a form of selection, and this explains in part why the general incidence is low. However, if the unilateral cryptorchid participates in breeding because the animal is outstanding, the tendency will be to concentrate the polygenes. Possibly the excellence of the stud may be improved, but at the price of a higher number of cryptorchids in later generations. Ideally, no unilateral cryptorchids should be used for breeding, however good they may be. Also it is unwise to breed from close relatives of cryptorchids if one is really concerned with the problem. Remember that this applies especially to sisters of the cryptorchid. The females are not able to suffer from cryptorchidism, but they can carry the polygenes for it and pass them to their sons.

Diabetes mellitus

This is an infantile form of the disease since affected dogs will begin to exhibit symptoms between the ages of 2 and 6 months of age. Body weight was normal at first but the rate of growth declined until the weight was only one half to two-thirds of normal. Water consumption increased, accompanied by frequent voiding of sweet smelling urine. Bilateral cataracts developed unless prevented by insulin therapy. The pancreas either lacked or was greatly deficient of islet B cells although A cells were present. The progression of the disease could be alleviated by the appropriate insulin treatment. The affected dogs belonged to the Keeshond breed. A genetic analysis indicated that the anomaly was inherited as a recessive and has been designated *dm* (Kramer *et al.*, 1980, 1988).

Diaphragmatic Hernia

A congenital failure of the diaphragmatic membrane to complete development has been described by Feldman *et al.*, (1968). The herniation was extensive and portions of the abdominal organs protruded into the thoracic cavity. The numbers of normal and affected pups approximated a 3:1 ratio in two litters of the same parents. Despite this, the assumption of recessive monogenic determination could only be accepted with caution because of the small numbers involved. However, a much stronger case for recessive inheritance for a similar it not identical anomaly has been presented by Valentine *et al.* (1988) in the Golden Retriever. The gene is symbolized by *dh*.

Esophageal Dilation

This defect cause persistent vomiting in young puppies as soon as they commence eating solid food. There is weight loss and dehydration. With a nutritious semi-solid diet, vomiting is less and growth is more normal. Osborne *et al.* (1967) showed that the condition is familial and probably produced by a recessive gene. The present cases were observed in the Wire-haired Fox Terrier.

A genetic study of the anomaly in pups of the Miniature Schnauzer dog unfortunately could not reach definite conclusions for the mode of inheritance (Cox *et al.* 1980). It seems probable that the anomaly did not manifest in a number of individuals. A high mortality of the pups added to the confusion. By 4–6 months of age, most of the young dogs showed a recovery from vomiting.

Extrahepatic Portosystemic Vascular Shunt

The defect is an abnormal connection between portal and systemic hepatic veins. The clinical symptoms are partial anorexia, loss of weight, intermittent vomiting and diarrhoea. The malady was observed in four young adult bitches of an inbred family of American Cocker Spaniels and, because of the close relationship, it was considered that the condition could be inherited (Rand *et al.*, 1988).

Gingival Hypertrophy

The defect is characterized by overgrowth of the gums. Burstone *et al.* (1952) observed several cases in a family of Boxers. Thus, the anomaly could be said to be familial and, if monogenically inherited, probably as a recessive.

Glossopharyngeal Defect

The most obvious sign of this anomaly is a narrowing of the tongue, particularly towards the tip, where the margins may appear to be folded inwards. This produces a narrow tongue, hence the popular name of "bird tongue". The anomaly is more profound than this, however, for the affected pups have no interest in suckling and indeed appear to lack the ability. They cannot swallow and cannot be taught to do so. As a consequence, the pups die of starvation. The anomaly is due to a recessive gene *bt* (Hutt and Lahunta, 1971).

High Uric Acid Excretion

The majority of dogs excrete small quantities of uric acid in the urine. However, the Dalmatian is exceptional in that these excrete large amounts. This does not appear to be detrimental to the individual and should be regarded as a peculiarity rather than an anomaly. Dalmatians do not appear to suffer more from kidney or bladder stones than other breeds, although the composition of such stones as found appears to differ. The difference in uric acid excretion is due to a recessive gene *ua* for high excretion. The *ua* gene is independent of the extreme white p^w and ticking *T* genes carried by the Dalmatian and responsible for the unique colouring of the breed. The *ua* gene has probably been fixed in the breed by chance variation during its early evolvement (Trimble and Keeler, 1938).

Intestinal Malabsorption

The anomaly is characterized by a persistent diarrhoea which sets in at about 2–6 months of age and which fails to respond to treatment. This is accompanied by loss of weight, severe emaciaton and muscular weakness. The cause appeared to be failure to absorb nutrients through the intestinal membranes, especially proteins and probably essential amino acids. The condition is inevitably fatal (Breitschwerdt *et al.*, 1980). Pedigree data for a strain of Basenjis are suggestive of recessive heredity and the gene is symbolized by *im*.

Laryngeal Paralysis

Numerous cases of the defect were observed among Bouvier dogs aged about 4–6 months. Affected animals show spontaneously occurring laboured breathing and vomiting due to laryngeal spasm. The malady appears to reside in dysfunction of the laryngeal abductor muscle. At autopsy, these and other muscles, together with their innervation, showed degenerative changes. The spasms frequently resulted in death (Venker-Haagen, 1978, 1981). The anomaly is due to a dominant gene *La*. The same or a similar recessively inherited defect has been observed for the Siberian Husky (O'Brien and Hendriks, 1986).

Lethal Oedema

This abnormality is a true congenital lethal condition. Afflicted pups are born dead and are grossly oedematous. Fluid is present under the skin and in the abdominal and thoractic cavities. The liver is congested and necrotic. The genetic data could be more extensive but, even so, these are fully consistent with monogenic recessive heredity (Ladds *et al.*, 1971). The responsible gene is symbolized by *lo*.

Mononephrosis

The absence of one kidney is suspected of being inherited in a recessive manner in the Beagle (Fox, 1964). However, too few animals were examined for the mode of inheritance to be conclusively established. The situation is complicated by cystic inflammation of the remaining kidney.

Pancreatic Insufficiency

This defect is produced by a deficiency of the pancreatic hydrolases and manifests between the ages of 1–3 years. The affected animals become emaciated in spite of a voracious appetite. Coprophagia is common. The faeces are plentiful, pale in colour, greasy and rancid smelling. The dog remains lively. Monogenic heredity is indicated for the German Shepherd Dog and the defect would seem to be due to a recessive gene *pc* (Weber and Freudiger, 1978; Westermarck, 1980).

Renal Cystic Adenocarcinoma

Lium and Moe (1985) have described a familial incidence of multiple renal cysts in the German Shepherd Dog. The kidneys were enlarged and contained several large and numerous small solid and cystic tumours. In a number of animals, the tumors had spread to many other internal organs and tissues. All of the dogs had a large number of firm nodules on the skin all over the body but especially on the head, lower limbs and back. Some bitches had similar nodules on the walls of the uterus.

Renal Dysfunction

Some six different breeds of dog have been reported as possessing a familial renal disease, but whether these are inherited or, if so, what is the pattern of heredity, is unknown (Kelly, 1986). The typical symptoms are excessive thirst (polydipsia) and urination (polyuria), vomiting, diarrhoea and marked loss of weight. The more likely cases for a genetic causation are those described for the Norwegian Elkhound (Finco *et al.*, 1970; Finco, 1976) and for the Soft-coated Wheaten Terrier (Eriksen and Grondalen, 1984; Nash *et al.*, 1984). In the latter breed, particularly, there is some evidence that the anomaly could be due to a recessive gene. Additional data are awaited with interest.

Samoyed Hereditary Glomerulopathy

This disease is a steady degeneration of the kidneys which varies in severity between the sexes. Males with affliction are healthy until about

3 months of age, when they cease to grow and become steadily thinner. Death occurs between 8 and 15 months from renal failure. Afflicted females become similarly emaciated but they do not die (Jansen *et al.*, 1986). The gene may be symbolized by *Shg*.

This malady is of interest genetically since the causative gene is sex-linked. That is, the gene is borne on the X chromosome and the disease is typically carried by the bitch and passed on to 50 per cent of her sons and daughters. The effect on her sons is eventually lethal as described above, since they have the gene on their single *X* chromosome. The effect on the daughters is less because they have the gene on only one of their *X* chromosomes, and the heterozygous genotype protects them from the final ravages of renal failure. Although thinner than normal, they are able to breed and perpetuate the anomaly. The sex-linked heredity will ensure that they will produce 50 per cent affected male offspring even when mated to unrelated males.

Sex Reversal

Potential sex is determined at fertilization and normally individuals with *XY* chromosomes will develop into males and those with *XX* chromosomes will develop into females. Selden *et al.* (1978, 1984) and Meyers-Wallen and Patterson (1988) have presented details of a family of Cocker Spaniels in which numerous *XX* hermaphrodites and *XX* "males" were observed. The former possessed both female and male sex structures: most were sterile but two actually produced progeny when mated to normal *XY* males. The *XX* "males" were sterile. These abnormal dogs will be detected either by the sterility or by their curious genital anatomy. All had levels of the male *H-Y* antigen comparable to that found in normal males. Additionally, several normal females had low levels of the male *H-Y* antigen.

Selden *et al.* advance the hypothesis that the sex reversal is due to a gene *sxr* which, when homozygous, results in a level of *H-Y* antigen high enough to cause the *XX* embryonic gonad to develop either partially (hermaphroditism) or completely into a male form. When heterozygous, the gene produces a level of *H-Y* antigen which, while detectable, is insufficient to affect the *XX* embryonic gonads and these develop as in normal females.

The observations of Williamson (1979) for the occurrence of hermaphrodites in a kennel of Kerry Blue Terriers would be in agreement with the hypothesis of Selden *et al.* In fact, Williamson did offer a tentative explanation along these lines.

Tumours

A frequent cause of mortality among dogs is cancerous growths or tumours of various kinds. Indeed, these could rank third after the scourge of virus diseases. Several attempts have been made to establish a link between specific or related cancers and a breed disposition but with little positive success. These studies have involved researching veterinary records for a significant rate of occurrence of a cancer type in certain breeds. However, those associations which have been found are not always consistent from one study to the next. This casts doubt on the generality of the conclusions. It is quite possible for a breed to have a liability to develop a particular type of cancer but the link, if it exists, could be elusive. The majority of cancers occur in aged animals and arise as a failure of the genetic regulation of cell physiology, such as normal functioning or renewal. The topic has been ably (and critically) reviewed by Willis (1989).

Umbilical Hernia

Umbilical hernia is the protrusion of a small piece of intestine through the region of the umbilicus; it usually follows from partial failure of the umbilical ring to close. The hernia is expressed as a soft lump in the middle of the abdomen. The incidence in the general population is low, but it may vary between breeds, a *prima facie* indication that a genetic factor is operating (Hayes, 1974).

A study of the defect in Cocker Spaniels by Phillips and Felton (1939) revealed that their cases appeared to be inherited as a monogenic recessive trait. This result is a little surprising, since an anomaly of this nature is commonly regarded as being a polygenic threshold character. The Phillips and Felton observations may prove to be exceptional.

Immunogenetics and Biochemical Genetics

These are two branches of genetics which are rapidly expanding at this time. The results being obtained are in intense interest to geneticists and to some sectors of veterinary medicine. In general, however, the results are unlikely to be of interest to dog breeders. For those who do wish to study these matters in depth, the reviews of Rapaport and Bachvaroff (1978), Vriesendorf (1979) and Colling and Saison (1980) should be read.

Although apparently removed from ordinary dog breeding, the genetic variation being uncovered by these studies can be useful in several respects. Most biochemical differences, such as for proteins or enzymes, are either harmless or have trivial effects, but others can herald serious deficiency diseases. Several of these latter, in fact, have been discussed earlier under the malady for which they are responsible.

A number of different blood groups have been discovered. At present none of these give rise to problems of blood transfusions, except perhaps in those cases where repeated transfusions are necessary. Differences of transplant antigens have been isolated and appear to be as complex as those

TABLE 19

Genetic Abnormalities and Their Symbols

Symbol	Name	Page
ac	Achondroplasia	216
ad	Acrodermatitis	201
al	Alopecia	202
am	Ataxia demyelination	229
an	Anury	227
at	Ataxia	229
bd	Brachydactyly	217
be	Bithoratic ectromelia	217
bh	Black hair follicle dysplasia	202
br	Brachyury	227
bt	Bird tongue	248
cb	Cerebellar ataxia	229
cbd	Cerebellar degeneration	230
cbh	Cerebellar ataxia hydrocephalous	230
cca	Cerebellar cortex atrophy	230
Cd	Connective tissue dysplasia	202
cea	Collie eye anomaly	207
cmo	Craniomandibular osteopathy	218
cn	Cyclic neutropenia	240
cp	Cleft palate	218
cr	Crystaline cornea dystrophy	208
crs	Cranioschisis	226
cs	Carpal subluxation	217
csp	Cervical spondylopathy	218
ct	Copper toxicosis	248
dan	Dwarf-anaemia	219
dea	Dobermann eye anomaly	208
dh	Diaphragmatic hernia	247
dm	Diabetes mellitus	247
Dmy	Dermatomyositis	203
Dpt	Dysprothrombinaenia	243
ds	Dermoid sinus	203
dw	Pituitary dwarfism	225
dwd	Dwarf, retina dysplasia	219
dyx	Progressive muscular dystrophy	235
ell	Elliptocytosis	240
es	Elbow subluxation	220
fu	Fucosidosis	231
ga1	GM1 gangliosidosis	232
ga2	GM2 gangliosidosis	232
gan	Giant axonal neuropathy	232
gc	Globoid-cell leukodystrophy	232
gl	Glaucoma	209
gr	Granulocytopathy	241
ha	Hairless, American	204
hc	Hyproproconvertaemia	243
he	Hemeralopia	210
Hma	Haemophilia A	243
Hmb	Haemophilia B	243
Hf	Hypofibrinogenaemia	243
hn	Hypertrophy neuropathy	233

TABLE 19 *continued*

Symbol	Name	Page
Hr	Hairless	204
im	Intestinal malabsorption	249
la	Laryngeal paralysis	249
ld	Bjerkas' leukodystrophy	229
li	Neuronal ceroid-hypfuseinosis	236
lo	Lethal oedema	249
Ly	Congenital lymphoedemia	239
lx	Lens luxation	210
mfd	Muscle fibre deficiency	233
mg	Myasthenia gravis	233
mp	Myelopathic paralysis	234
mrd	Multifocal retinal dysplasia	211
my	Sex-linked myopathia	238
na	Neoronal abiotrophy	236
nc	Narcocataplexy	234
nma	Neurogenic muscular atrophy	235
pa	Progressive axonopathy	236
pc	Pancreatic atrophy	250
pd	Perthe's disease	224
pfk	PFK deficiency	241
pg	Pelger-Huet anomaly	244
pk	PK deficiency	241
pra	Progressive retina atrophy	212
Pta	Plasma thromboplastic antecedent deficiency	243
qu	Quadriplegia	237
rd	Retina dysplasia	214
ro	Rod-cone dysplasia	214
sc	Scottie cramp	237
sh	Shaking	238
Shg	Samoyad hereditary glomerulopathy	250
Sl	Skeleton lethal	226
sm	Short mandible	223
sma	Spinal muscular atrophy	239
sp	Short spine	226
Stf	Stuart-Prower factor	243
sxr	Sex reversal	251
td	Tapetal degeneration	215
tr	Thoratic hemivertebrae	227
ua	High uric acid excretion	249
Wvd	Von Willebrand's disease	243

of other mammals. This implies that complications of matching donor and host will arise should organ transplants ever be contemplated for dogs.

A spin-off of this sort of research is that individual dogs can be typed for biochemical variation and for the various transplant and blood antigens to an extent hitherto impossible. One outcome is that the results of matings can be checked with a fair degree of accuracy. For example, the mother of a litter is rarely unknown, but the father may be in doubt. An analysis of the blood and serum of the mother and pups could indicate which of a number of stud dogs could be the father. It is not possible to say that a given dog is the sire of a litter, but it is possible to say that he is not. Cases of mistaken or disputed paternity could be decided in principle by such an analysis. The greater the number of biochemical differences which can be identified, the greater the likelihood of positive discrimination between suspects. However, the procedures are exacting and specialized, but a laboratory could probably be set up to handle these if the demand for its services were sufficiently large. Vriesendorp *et al.* (1974) and Juneja *et al.* (1987) have clarified a number of cases of disputed or questionable paternity by the methods outlined above.

Recently, a technique has been developed by which tiny but highly variable sections of the chromosomes can be portrayed as a series of dark bars on a photographic film (Jeffreys and Morton, 1987; Georges *et al.*, 1988*a*,*b*). The significant property of the bars is that these are unique to the individual. This has led to the technique being termed "genetic fingerprinting". The series of bars resemble the identifying computer codes to be found on items of merchandise and for this reason have come to be known as genetic bar codes. Furthermore, the individual bars are inherited from parent to offspring in a consistent manner. The uniqueness of the bar codes for the individual and their genetic descent implies that these have immediate practical applications. They may be used to decide more decisively than hitherto cases of uncertain or disputed paternity. Instead of simply indicating that a certain stud dog could not have sired a specific pup, it is now possible to be more positive and state that only a certain dog could be the sire. Genetic fingerprinting is available as a commercial service. The usual procedure is to take small samples of blood from the dog(s) under test, by arrangement with a vet, for submission to the laboratory. Full details of the service for breeders may be obtained from vets.

9.

Behaviour Aspects

It is not proposed to delve deeply into dog behaviour, not because the topic is devoid of interest but because, out of the considerable literature, little has been written on the genetic differences separating the breeds. That inherited differences do exist is indisputable, as shown by the various breed categories with different modes of behaviour. A small part of this variation could be imitative or learnt from the mother, siblings or older members of a kennel, but the major part is instinctive. That is, the behavioural pattern or patterns are genetic. This is not to say that canine behaviour is stereotypic, for the dog has remarkable powers of learning, but the ability of the animal to perceive its surroundings, react to mishaps, respond to appropriate training and accept man as a companion and leader, is controlled genetically to a significant extent. An excellent summary of the early studies on the genetics of behaviour has been provided by Burns and Fraser (1966).

In almost all serious studies of behaviour, the heritable aspect has been found to be polygenic in nature. Some of the early investigators have indeed attempted to interpret variation of certain attributes to the assortment of major genes, e.g. hunting or trail barking versus non-barking, hunting with head held high versus head held low, or gun-shyness versus non-gun-shyness, but this was in a period of genetics where simple Mendelian inheritance was too readily adopted. Nowadays, such variation would be sharply scrutinized before a monogenic interpretation is proposed. This is not to say that major genes could not be involved, but that their presence should not be uncritically accepted.

As Scott and Fuller (1965) have pointed out, the dog is not a super-wolf. In those instances where the dog has successfully returned to the wild, this has only been achieved for areas in which the wolf is absent, e.g. Australia, as the dingo. These authors depict the dog population as a group of specialists. In almost any aspect of behaviour, a breed or breeds exist which surpass the wolf in ability. The Greyhound type dogs are more fleet of foot than the wolf, even if this is accomplished at the expense of heavy muscles which serve a wolf well. Tracking breeds are probably superior to wolves in detection and persistency in following a trail. Sheepdogs are superior herders and are adapted to work with man. Wolves are said to take little interest in birds, but many dog breeds excell in this respect (pointing). This latter feature, however, is an ability which is only useful in co-operation with man. The point of listing this diversity is to emphasize that man has modified the behaviour of the dog as adroitly as purely physical traits. The latter can be seen by inspection, whereas the former can only be appreciated after intimate contact with different breeds.

The breed differences of behaviour have come into being as the result of selective breeding of dogs for specific purposes. It is necessary not only to change the physical characteristics but also the behaviour. For example, guard dogs are not only required to be large in stature but also to be alert and have some degree of aggressiveness. Hunting dogs are not only to be fast runners but also interested in finding game. Shepherd dogs must not only be capable of herding sheep but also be responsive to the shepherd in driving the flock. Tracking dogs must not only be able to make fine discrimination of scents but possess tenacity of purpose. For some police work, such as sniffing out illegal drugs or contraband goods, fine discrimination is probably sufficient.

Over the millennia, man has selected and perfected breeds with remarkable, almost uncanny, abilities. In the beginning, breeders probably made use of variation of behaviour shown by individual animals, but by mating these together he began to select for the genetic background to the traits in which they were interested. In this manner, the breeds differentiated and some became capable of greater feats. No doubt the training methods improved as the dogs became increasingly more responsive to intricate instructions. The ability to recognize, learn, remember and obey human commands is an aspect of canine intelligence. It is the domestic counterpart of the wolf's ability to recognize and act upon cues signalled by other members of a pack. However, it is doubtful if the wolf has the capability of the better working dog breeds of today.

The analysis of the aptitudes which distinguishes the ability of working dogs has been discussed by several authors. Whitney (1947) makes a distinction between aptitudes which are innate to the dog and those which are learnt or brought out by training. The former can also be enhanced by proper training. In crosses between breeds which displayed the aptitudes either poorly or well, the F_1 often showed a predominance of one aptitude, but in subsequent generations the inheritance gave indications of being polygenic. Although it may be argued that Whitney's observations were not well controlled, his experience as a dog breeder and veterinarian should not be discounted for this sort of study, where a close relationship between man and dog is useful in evaluating the results.

For some 20 years, Scott and Fuller (1965) have been studying the behaviour of five breeds: Basenji, Beagle, Cocker Spaniel, Shetland Sheepdog and Wire-haired Fox Terrier. These breeds were chosen as representative of the main classes of dog. Additionally, crosses were made between the Basenji and Cocker Spaniel and F_2 and backcross generations were obtained. All five breeds and the crosses were subjected to a battery of tests designed to detect differences of sociability, temperament and learning. These tests are the most extensive and well controlled so far performed with dogs. The results are described and discussed at length in the above report.

A significant finding was that of critical periods in the development of the puppy. A critical period is defined as the stage when a small exposure to stimuli will have a profound effect on later behaviour. The various tests suggest the following periods and stages:

Neonatal	1–2 weeks
Transitional	2–3 weeks
Socialization	3–9 weeks
Optimum socialization	5–7 weeks
Playful fighting commences	3–5 weeks (average)
Tail wagging commences	4 weeks (average)
Fear response commences	5 weeks

During the neonatal period, the pup is growing steadily but is virtually passive, showing little response to stimuli. However, between 2 and 3 weeks, a rapid change is evident; the eyes open and the pup is responsive to sounds. The pup begins to adapt to its environment. The learning and socialization period lasts for some weeks and is important for the young pup. If a pup does not have contact with other dogs and humans during the period, it behaves strangely and fearfully towards them in later life. Furthermore, there appears to be an interval of optimum socialization. A pup whose first contact with man is made between the ages of 5-7 weeks socializes more quickly than if the contact is made before or after this age span. It is intriguing to have such a narrow age range during which socialization occurs, whether one considers the whole period or the optimum part of it. The socialization process appeared to be akin to imprinting of birds, but is probably not so inviolable. There appeared to be little difference in socialization between the five breeds, except that the Basenji initially stood a little apart from the others for some features, but soon caught up with the others.

The major breed differences occurred for the tests of temperament and of learning ability. There is little point in attempting to paraphrase these, as they were many and varied. The reader would be advised to study the conclusions of Scott and Fuller for him or herself. It is interesting that most of the differences are statistically significant (as the statisticians say), that is, they have real meaning. This would confirm the impressions formed by most people who are familiar with the five breeds involved. The F_1, F_2 and backcross generations also revealed comparable differences. For most tests, a component of the variation could be isolated which is due to heredity. However, the size of the genetic component of the variation varied from test to test, indicating that different problems called for different abilities which are not determined genetically to the same extent. Thus, the genetic background to a behavioural trait is not the same as that for a learning trait. This conclusion could be deduced on general grounds, but it is gratifying to have confirmatory experimental evidence.

One of the problem-solving tests was made steadily more difficult as these were given to the dogs. An interesting result became apparent as a consequence. If the problem was made difficult too soon, the animal would become discouraged. The dog would make attempts to solve the problem, but if several of these ended in failure it would lose interest and eventually make no effort. On the other hand, if the dog was successful in the simpler tests it would persist more strenuously in the more difficult, even if these were above its capability. Success in the form of rewards (tit-bits, say) for achievement would seem to be important for the canine psyche. Presumably successful dog trainers make full use of this method, never moving too quickly in the training so that the dog becomes discouraged, but continuing steadily until the animal has reached the limit of its powers. Burns and Fraser (1966) write of the method as giving the animal a high motivation to succeed. It should not be assumed that training is more important than the genetic potential of the breed. The two aspects are complementary. No amount of training can make a "dull" dog the equal of a "bright" dog. The most that can be expected is that the gap could be narrowed, but even this is uncertain if the bright dog also receives special attention.

Two books on working dogs, which are of unusual interest, are those of Humphris and Warner (1934) and Kelly (1949). These books discuss in detail the training and breeding of dogs, the first in the more general sense and the second for Australian shepherding. These authors made use of genetic principles in carefully scoring the dogs for specific aptitudes, selection of the better scoring individuals, ensuring that these latter contributed substantially to the gene pool and, in some cases, progeny testing. The results obtained by these methods were highly satisfactory.

An important area of dog training and performance is that of providing guide dogs for the blind. Reliable techniques of breeding and training for this purpose is clearly of some significance. One of the most informative works on the subject is that of Pfaffenberger (1963). The majority of guide dogs are usually drawn from a select group of breeds which are known to yield animals of suitable temperament and trainability. The proportion of dogs which can benefit from the rigorous training may vary according to breed, but cannot be very high. A programme was launched not merely to train suitable guide dogs but to relate excellence in the adult dog to special aptitudes shown in the puppy stage. Furthermore, selection of breeding stock was based upon guide dog performance, ignoring exhibition or breed points The results were remarkable. In one strain of German Shepherd Dogs the proportion of successful dogs rose from 9 per cent to 90 per cent in about 12 years. Evidently much of the variability in aptitudes which make a good guide dog is inherited and hitherto had not been concentrated in special "guide dog" breeds.

In the present case success was achieved by careful attention to grading and scoring the pup for pertinent traits through to maturity, and using this knowledge in the selective breeding. It is reasonably certain that this programme would not have been successful but for the insights of dog behaviour which were gained from the Jackson Laboratory observations. No real efforts were made to determine the mode of inheritance of the attributes fundamental for a good guide dog. This was not the point of the programme. The amount of progress observed for some traits, however, was such to suggest that these were determined by a few, rather than many, genes. Expressions obtained in this manner can be deceptive, since much will depend upon the effectiveness of the selection procedures. These seemed to be efficient in the present case, a fact which could explain their effectiveness.

It should not be thought that selection for behaviour characteristics can be relaxed. Certainly not, and in the compilation of a total score formula items should be included for behaviour. Certain of these items must relate to those characters which make the breed what it is. However, there are a few traits which are undesirable for most breeds, and selection should be directed against these. For example, breeds bred as companions or exhibition should be tractable and not display aggressiveness or viciousness of any kind. This trait has been bred out of most dogs, but a few do show signs of not being wholly trustworthy. Perhaps this is due to mismanagement (as a puppy or whatever), but nonetheless such dogs are undesirable. Also, there is the excessively excitable, nervous or "highly strung" animal. This sort of behaviour is inexcusable. Although one can argue that this is due to mismanagement, such behaviour is likely to be inherited and individuals showing such behaviour should not be used for breeding.

References

Aguirre, G., Farber, D., Lolley, R., Fletcher, R. T., and Chader, G. J. (1978) Rod-cone dysplasia in Irish Setters: a defect in cyclic GMP metabolism in visual cells. *Science*, **201**: 1113–1134.

Aguirre, G. D., and Rubin, L. F. (1972) Progressive retinal atrophy in the Miniature Poodle. *J. Amer. Vet. Med. Assoc.*, **160**: 191–201.

Alonso, R. A., Hernandez, A., Diaz, P., and Cantu, J. M. (1982) An autosomal recessive form of hemimelia in dogs. *Vet. Rec.*, **110**: 128–129

Alroy, J., Orgad, U., Ucci, A. A., Schelling, S. H., Schunk, K. L., Warren, C. D., Raghavan, S. S., Kolodny, E. H. (1985) Neurovisceral and skeletal GMl gangliosidosis in dogs. *Science*, **229**: 470–472

Andersen, S. A., Andersen, E., and Christensen, K. (1988) Hip-dysplasia selection index exemplified by data from German Shepherd dogs. *J. Anim. Breed. Genet.*, **105**: 112–119.

Anderson, M., Henricson, B., Lundquist, P. G., Wedenberg, and Wersall, J. (1968) Genetic hearing impairment in the Dalmatian dog. *Acta Otolaryngol. Suppl.*, **232**: 1–34.

Andresen, E. (1977) Haemolytic anaemia in Basenji dogs. *Anim. Blood. Grps. Biochem. Genet.*, **8**: 149–156.

Andresen, E., and Willenberg, P. (1976a) Pituitary dwarfism in Carelian Bear dogs. *Hereditas*, **84**: 232–234.

Andresen, E., and Willenberg, P. (1976b) Pituitary dwarfism in German Shepherd dogs. *Nord. Vet. Med.*,**28**: 481–486.

Appleby, E. C., Longstaffe, J. A., and Bell, F. R. (1982) Ceroid-lipofuscinosis in two Saluki dogs. *J. Comp. Pathol.*, **92**: 375–380.

Ashton, N., Barnett, K. C., and Sachs, D. D. (1968) Retina dysplasia in the Sealyham Terrier. *J. Pathol. Bacteriol.*, **96**: 269–272.

Averill, D. R., and Bronson, R. T. (1977) Inherited necrotizing myelopathy of Afghan Hounds. *J. Neuropath. Exp. Neurol.*, **36**: 734–737.

Baker, T. L., Foutz, A. S., McNerney, V., Mitler, M. M., and Dement, W. C. (1982) Canine model of narcolepsy. *Exper. Neurol.*, **75**: 729–742.

Ball, M. U., McGuire, J. A., Swain, S. F., and Hoerlein, B. F. (1982) Patterns of occurrence of disk disease among Dachshunds. *J. Amer. Vet. Med. Assoc.*, **180**: 519–522.

Bardens, J. W., Bardens, G. W., and Bardens, B. (1961) Clinical observations on a von Gierke-like syndrome in puppies. *Applied Vet.*, **32**: 4–7.

Bargai, U., Waner, T., and Beck, Y. (1988) Canine hip dysplasia in Israel: sixteen years of a control scheme. *Israel J. Vet. Med.*, **44**: 202–207.

Barker, C. G., Herrtage, M. E., Shanahan, F., and Winchester, B. G. (1988) Fucosidosis in English Cocker Spaniels: results of a trial screening programme. *J. Small Anim. Pract.*, **29**: 623–630.

Barnett, K. C. (1969) Primary retinal dystrophies in the dog. *J. Amer. Vet. Med. Assoc.*, **154**: 804–808.

Barnett, K. C. (1976) Comparative aspects of canine hereditary eye disease. *Advance. Vet. Sci. Comp. Med.*, **20**: 39–67.

Barnett, K. C. (1978) Hereditary cataract in the dog. *J. Small Anim. Pract.*, **19**: 109–120.
Barnett, K. C. (1980) Hereditary cataract in the Welsh Springer Spaniel. *J. Small Anim. Pract.*, **21**: 621–625.
Barnett, K. C. (1985a) Hereditary cataract in the Miniature Schnauzer. *J. Small Anim. Pract.*, **26**: 635–644.
Barnett, K. C. (1985b) The diagnosis and differential diagnosis of cataract in the dog. *J. Small Anim. Pract.*, **26**: 305–316.
Barnett, K. C. (1986) Hereditary cataract in the German Shepherd dog. *J. Small Anim. Pract.*, **27**: 387–395.
Barnett, K. C. (1988) Inherited eye disease in the dog and cat. *J. Small Anim. Pract.*, **29**: 462–475.
Barnett, K. C., Bjorck, G. R., and Koch, E. (1970) Hereditary retina dysplasia in the Labrador Retriever in England and Sweden. *J. Small Anim. Pract.*, **10**: 755–759.
Barnett, K. C., and Startup, F. G. (1985) Hereditary cataract in the standard poodle. *Vet. Rec.*, **117**: 15–16.
Bedford, P. G. C. (1980) The aetiology of canine glaucoma. *Vet. Rec.*, **107**: 76–82.
Bedford, P. G. C. (1982a) Collie eye anomaly in the United Kingdom. *Vet. Rec.*, **111**: 263–270.
Bedford, P. G. C. (1982b) *Vet. Rec.*, **111**: 304.
Bedford, P. G. C. (1984a) Retinal pigment epithelial dystrophy: a study of the disease in the Briard. *J. Small Anim. Pract.*, **25**: 129–138.
Bedford, P. G. C. (1984b) Retina dysplasia in the dog. *Vet. Annual,* **24**: 325–328.
Bedford, P. G. C. (1989) Control of inherited retinal degeneration in dogs. *J. Small Anim. Pract.*, **30**: 172–177.
Bellhorn, R. W., Bellhorn, M. B., Swarm, R. L., and Impellizzeri, G. W. (1975) Hereditary tapetal abnormality in the Beagle. *Ophthalmic Res.*, **7**: 250–260.
Bergsjo, T., Arnesen, K., Heim, and Nes, N. (1984) Congenital blindness with ocular development anomalies, including retinal dysplasia, in Doberman Pinschers. *J. Amer. Vet. Med. Assoc.*, **184**: 1383–1386.
Birns, M. S., Bellhorn, R. W., Impellizzeri, C. W., Aguirre, G. D., and Laties, A. M. (1988a) Development of hereditary tapetal degeneration in the Beagle dog. *Current Eye Res.*, **7**: 103–114.
Bjerkas, I. (1977) Hereditary "cavitating" leukodystrophy in Dalmatian dogs. *Acta Neuropathol.*, **40**: 163–169.
Bjorck, G., Mar, W., Olsson, S. G., and Sourander, P. (1962) Hereditary ataxia in Fox Terriers. *Acta Neuropathol.,* Suppl. **1**: 45–58.
Bogart, R. (1959) *Improvement of Livestock,* MacMillan.
Boom, H. P. A. (1965) Anomalous animals. *S. Afr. J. Sci.*, **61**: 159–171.
Bowles, C. A., Alsaker, R. D., and Wolfe, T. L. (1979) Studies of the Pelget-Huet anomaly in Foxhounds. *Amer. J. Pathol.*, **96**: 237–248.
Breitschwerdt, E. B., Halliwell, W. H., Foley, C. W., Start, D. R., and Corwin, L. A. (1980) A hereditary diarrhetic syndrome in the Basenji characterized by malabsorption, protein losing enteropathy and hypergammaglobulinemia. *J. Amer. Anim. Hosp. Assoc.*, **16**: 551–560.
Briggs, L. C. (1940) Some experimental matings of colour bred White Bull Terriers. *J. Hered.*, **31**: 236–238.
Briggs, L. C., and Kaliss, N. (1942) Coat colour inheritance in Bull Terriers. *J. Hered.*, **33**: 222–228.
Briggs, O. M. (1985) Lentiginosis profusa in the pug. *J. Small Anim. Pract.*, **26**: 675–680.
Brown, R. V., and Teng, T. S. (1975) Studies of inherited pyruvate kinase in the Basenji. *J. Amer. Anim. Hosp. Assoc.*, **11**: 362–365.
Burns, M. (1943) Hair pigmentation and the genetics of colour in Greyhounds. *Proc. Roy. Soc. Edin. B.*, **61**: 462–490.
Burns, M., and Faser, M. N. (1966) *Genetics of the Dog.* Oliver & Boyd.
Burns, M. S., Tyler, N. K., and Bellhorn, R. W. (1988b) Melasome abnormalities of ocular pigmented epithelial cells in Beagle dogs with hereditary tapetal degeneration. *Current Eye Res.*, 7: 115–123.
Burstone, M. S., Bond, E., and Litt, R. (1952) Familial gingival hypertrophy in the dog. *Arch. Pathol.*, **54**: 208–212.

Cambell, K. L. (1985) Canine cyclic hematopoiesis. *Comp. Cont. Educ. Pract. Vet.*, **7**: 57–60.
Carmichael, S., Griffiths, I. R., Harvey, M. J. A. (1983) Familial cerebellar ataxia with hydrocephalus in Bull Mastiffs. *Vet. Rec.*, **112**: 354–358.
Carrig, C. B., and Seawright, A. A. (1969) Familial canine polyostotic fibrous dysplasia with subperiosteal cortical defects. *J. Small Anim. Pract.*, **10**: 397–405.
Carrig, C. B., MacMillan, A., and Brundage, J. (1977) Retinal dysplasia associated with skeletal abnormalities in Labrador Retrievers. *J. Amer. Vet. Med. Assoc.*, **170**: 49–57.
Carrig, C. B., Sponenberg, D. P., Schmidt, G. M., and Tvedten, H. W. (1988) Inheritance of associated ocular and skeletal dysplasia in Labrador Retrievers. *J. Amer. Vet. Med. Assoc.*, **193**: 1269–1272.
Carver, E. A. (1984) Coat colour genetics of the German Shepherd dog. *J. Hered.*, **75**: 247–252.
Cheville, N. F. (1968) The Gray Collie syndrome. *J. Amer. Vet. Med. Assoc.*, **152**: 620–630.
Clark, R. G., Hartley, W. J., Burgess, G. S., Cameron, G. S., and Mitchell, G. (1982) Suspected inherited cerebellar neuroaxonal dystrophy in Collie Sheep dogs. *N.Z. Vet. J.*, **30**: 102–103.
Clough, E., Pyle, R. L., Hare, W. C. D., Kelly, D. F., and Patterson, D. F. (1970) An XXY sex chromosome constitution in a dog with testicular hypoplasia. *Cytogenetics.* **9**: 71–77.
Colling, D. T., and Saison, R. (1980) Canine blood groups. *Anim. Blood Grps. Biochem. Genet.*, **11**: 1–12.
Comfort, A. (1960) Longevity and mortality in dogs of four breeds. *J. Geront.*, **15**: 126–129.
Comfort, A. (1956) Longevity and mortality of Irish wolfhounds. *Proc. Zool. Soc. Lond.*, **127**: 27–34.
Conaway, D. H., Padgett, G. A., Bunton, T. E., Nachreiner, R. and Hauptman, J. (1985) Clinical and histological features of primary progressive thyroiditis in a colony of Borzoi dogs. *Vet. Pathol.*, **22**: 439–446.
Conroy, J. D., Rasmusen, B. A., and Small, B. (1975) Hypotrichosis in Miniature Poodle siblings. *J. Amer. Vet. Assoc.*, **166**: 697–699.
Cooper, B. J., Lahunter, A., Cummings, J. F., Lein, D. H., and Karrema, G. (1984) Canine inherited hypertrophic neuropathy. *Amer. J. Vet. Res.*, 1172–1177.
Cooper, H. J., Winaand, N. J., Stedman, H., et al. (1988) The homologue of the Duchene locus is defective in X linked muscular dystrophy of dogs. *Nature*, **334**: 154–156.
Cork, L. C., Griffin, J. W., Munnell, J. F., Lorenz, M. D., Adams, R. J., and Price, D. L. (1979) Hereditary canine spinal muscular atrophy. *J. Neuropath. Exp. Neurol.*, **38**: 209–221.
Cork, L. C., Troncoso, J. C., Price, D. L. (1981) Canine inherited ataxia. *Ann. Neurol.*, **9**: 492–499.
Cork, L. C., Troncoso, J. C., Price, D. L., Stanley, E. F., and Griffin, J. W. (1983) Canine neuroaxonal dystrophy. *J. Neuropath. Exper. Neurol.*, **42**: 286–296.
Corley, E. A., Sutherland, T. M., and Carlson, W. D. (1968) Genetic aspects of canine elbow dysplasia. *J. Amer. Vet. Med. Assoc.*, **153**: 543–547.
Cottrell, B. O., and Barnett, K. C. (1988) Primary glaucoma in the Welsh springer spaniel. *J. Small Anim. Pract.*, **29**: 185–199.
Cox, V. S., Wallace, L. J., Anderson, V. E., and Rushmer, R. A. (1980) Hereditary esophageal dysfunction in the Miniature Schnauzer dog. *J. Vet. Res.*, **41**: 326–330.
Cummings, J. F., Cooper, B. J., Lahunter, A., and Winkle, T. J. (1981a) Canine inherited hypertrophic neuropathy. *Acta Neuropathol.*, **53**: 137–143.
Cummings, J. F., Lahunter, A., and Winn, S. S. (1981b) Acral mutilation and nociceptive loss in English Pointer dogs. *Acta Neuropathol.*, **53**: 119–127.
Cunningham, J. G., and Farnbach, G. C. (1988) Inheritance and idopathic canine epilepsy. *J. Amer. Anim. Hosp. Assoc.*, **24**: 421–424.
Curtis, R. (1982) Primary hereditary cataract in the dog. *Vet. Annual*, **22**: 311–318.
Curtis, R. (1984) Late onset cataract in the Boston terrier. *Vet. Rec.*, **115**: 577–578.
Curtis, R., and Barnett, K. C. (1980) Primary lens luxation in the dog. *J. Small Anim. Pract.*, **21**: 657–668.
Curtis, R., and Barnett, K. C. (1989) A survey of cataracts in golden and Labrador retrievers. *J. Small Anim. Pract.*, **30**: 277–286.
Curtis, R., Barnett, K. C., and Leon, A. (1984) Persistent hyperplastic primary vitreous in the Staffordshire Bull Terrier. *Vet. Rec.*, **115**: 385.

Curtis, R., Barnett, K. C., and Startup, F. G. (1983) Primary lens luxation in the Miniature Bull Terrier. *Vet. Rec.*, **112**: 328–329.
Curtis, R. L., English, C., and Kim, Y.J. (1964) Spina bifida in a "stub" dog stock, selectively bred for short tails. *Anat. Rec.*, **148**: 365
Dain, A. R., and Walker, R. G. (1979) Two intersex dogs with mosaicism. *J. Reproduct. Fert.*, **56**: 239–242.
Dausch, D., Wegner, W., Michaelis, M., and Reetz, I. (1977) Ophthalmologische befunde in einer merlezucht. *Dstche Tierarztl. Wschr.*, **84**: 469–475.
Davis, S. J. M., and Valla, F. R. (1978) Evidence for domestication of the dog 12,000 years ago in Natufian of Israel. *Nature*, **276**: 608–610.
Dawson, W. M. (1937) Heredity in the dog. *U.S. Depart. Agric. Yearbook, 1937*: 1314–1343.
Desnick, R. J., McGovern, M. M., Schuchman, E. H., and Haskins, M. E. (1982) Animal analogues of human metabolic diseases. In Desnick, R.J. (Editor) *Animal Models of Inherited Metabolic Diseases*. Alan R. Liss, Inc.
Dice, P. F. (1980) Progressive retina atrophy in the Samoyed. *Mod. Vet. Pract.*, **61**: 59–60.
Dodds, W. J. (1974) Hereditary and acquired hemorrhagic disorders in animals. In Spaet, T. H. (Editor) *Progress in Hemostasis and Thrombosis*. Grune & Stratton.
Dodds, W. J. (1988) Third international registry of animal models of thrombosis and hemorrhagic diseases. *ILAR News*, **30**: R1–R32.
Draper, D. D., Kluge, J. P., and Miller, W. J. (1976) Clinical and pathologic aspects of spinal dysraphism in dogs. *Proc. XX World Vet. Congr.*, **1**: 134–137.
Druckneis, H. (1935) [Sex ratio and litter size in the dog.] *Vet. Med. Diss. Unit. Munchen.*
Duncan, I. D., and Griffiths, I. R. (1981) Canine giant axonal neuropathy. *J. Small Anim. Pract.*, **22**: 491–500.
Duncan, I. D., and Griffiths, I. R. (1982) A sensory neuropathy affecting long haired Dachshund dogs. *J. Small Anim. Pract.*, **23**: 381–390.
Edmonds, H. L., Hegreberg, G. A., Gelder, N. M., Sylvester, D. M., Clemmons, R. M., and Chatburn, C. G. (1979) Spontaneous convulsions in Beagle dogs. *Fed. Proc.*, **38**: 2424–2428.
Eigenmann, J. E., Zanesco, S., Arnold, U., and Froesch, E. R. (1984) Growth hormone and insulin-like growth factor I in German shepherd dwarf dogs. *Acta Endocrinol.*, **105**: 289–293.
Epstein, H. (1971) *The Origin of Domestic Animals of Africa*. Africana Publishing Corporation.
Eriksen, K., and Grondalen, J. (1984) Famileal renal disease in soft coated wheaten terriers. *J. Small Anim. Pract.*, **25**: 489–500.
Falco, M. J., Barker, J., and Wallace, M. E. (1974) Genetics of epilepsy in the British Alsatian. *J. Small Anim. Pract.*, **15**: 685–692.
Farrow, B. R. H., and Malik, R. (1981) Hereditary myotonia in the Chow Chow. *J. Small Anim. Pract.*, **22**: 451–465.
Feldman, D. B., Bree, M. M., and Cohen, B. J. (1968) Congenital diaphragmatic hernia in neonatal dogs. *J. Amer. Vet. Med. Assoc.*, **153**: 942–944.
Fiennes, R., and Fiennes A. (1968) *The Natural History of the Dog*. Weidenfeld & Nicolson.
Finco, D. R. (1976) Familial renal disease in Norwegian Elkhound dogs. *Amer. J. Vet. Res.*, **37**: 87–91.
Finco, D. R., Kurtz, H. J., Low, D. G., and Perman, V. (1970) Familial renal disease in Norwegian Elkhound dogs. *J. Amer. Vet. Med. Assoc.*, **156**: 747–760.
Flagstad, A. (1982) A new hereditary neuromuscular in the dog. *Hereditas*, **96**: 211–214.
Fletch, S. M., Pinkerton, P. H., and Brueckner, P. J. (1975) The Alaskan Malamute chrondrodysplasia (dwarfism-anemia) syndrome. *J. Amer. Anim. Hosp. Assoc.*, **11**: 353–361.
Fogh, J. M., and Fogh, I. T. (1988) Inherited coagulation disorders. *Vet. Clin. N. Amer.*, **18**: 231–243.
Foley, C. W., Lasley, J. F., and Osweiler, G. D. (1979) *Abnormalities of Companion Animals*. Iowa State Univ. Press.
Ford, L. (1969) Hereditary aspects of human and canine cyclic neutropenia. *J. Hered.*, **60**: 293–299.
Fox, M. W. (1964) Inherited polycystic mononephrosis in the dog. *J. Hered.*, **55**: 29–30.
Fox, M. W. (1978) *The Dog: Its Domestication and Behaviour*. Garland STPM Press.
Frankling, E. (1971) *The Dalmatian*. Popular Dogs.

Fritz, T. E., Lombard, L. S., Tyler, S. A., and Norris, W. P. (1976) Pathology and familial incidence of orchitis and its relation to thyroiditis in a closed Beagle colony. *Exper. Molec. Pathol.*, **24**: 142–158.
Gage, E. D. (1975) Incidence of clinical disc disease in the dog. *J. Amer. Anim. Hosp. Assoc.*, **11**: 135–138.
Gardner, D. L. (1959) Familial canine chondrodystrophia foetalis. *J. Path. Bact.*, **77**: 243–247.
Garmer, L. (1986) Linsluxation hos tibetansk terrier. *Svensk Vet.*, **38**; 132–133.
Georges, M., Hilbert, P., Lequarre, S. S., Leclerc, V., Hanset, R., and Vassart, G. (1988a) Use of DNA bar codes to resolve a canine paternity dispute. *J. Amer. Vet. Med. Assoc.*, **193**: 1095–1098.
Georges, M., Lequarre, A. S., Castelli, M., Hanset, R., and Vassart, G. (1988b) DNA fingerprinting in domestic animals using four different minisatellite probes. *Cytogenet. Cell Genet.*, **47**: 127–131.
Gelatt, K. N. (1972) Familial glaucoma in the Beagle dog. *J. Amer. Anim. Hosp. Assoc.*, **8**: 23–28.
Gelatt, K. N., and Gum, G. (1981) Inheritance of primary glaucoma in the Beagle. *Amer. J. Vet. Res.*, **42**: 1691–1693.
Gelatt, K. N., Powell, N. G., and Huston, K. (1981) Inheritance of microphthalmia with coloboma in the Australian Shepherd dog. *Amer. J. Vet. Res.*, **42**: 1686–1690.
Gelatt, K. N., Samuelson, D. A., Bauer, J. E., Das, N. D., et al. (1983) Inheritance of congenital cataracts and microphthalmia in the miniature Schauzer. *Amer. J. Vet. Res.*, **44**: 1130–1132.
Gelatt, K. N., Whitley, R. D., Lavach, J. D., Barrie, K. P., and Williams, L. W. (1979) Cataract in Chesapeake Bay Retrievers. *J. Amer. Vet. Med. Assoc.*, **175**: 1176–1178.
Giger, U., and Harvey, J. W. (1987) Hemolysis caused by phosphofrucktokinase deficiency in English Springer Spaniels. *J. Amer. Vet. Med. Assoc.*, **191**: 453–459.
Giger, U., Reilly, M. P., Asakura, T., Baldwin, C. J., and Harvey, J. W. (1986) Autosomal recessive inherited phosphofructokinase deficiency in English Springer Spaniel dogs. *Anim. Genet.*, **17**: 15–23.
Green, E. L. (1957) Mutant stocks of cats and dogs offered for research. *J. Hered.*, **48**: 56–57.
Griffiths, I. R. (1986) Inherited neuropathies of dogs. *Vet. Annual*, **25**: 287–292.
Griffiths, I. R., Duncan, I. D., and Barker, J. (1980) Progressive axonopathy of Boxer dogs. *J. Small Anim. Pract.*, **21**: 29–43.
Griffiths, I. R., Duncan, I. D., McCulloch, M., and Harvey, M. J. A. (1981) Shaking pups: a disorder of the cent myelination in the Spaniel dog. *J. Neurol. Sci.*, **50**: 423–433.
Grondalen, J., and Lingaas, F. (1988) Arthrosis of the elbow joint among Rottweiler dogs. *Tid. Diergenk. (Suppl.)*, **113**: 49S–51S.
Grun, E. (1985) Erbliche blutgerinnungsdefekte bei haustieren. *Mh. Vet. Med.*, **40**: 746–749.
Gruneberg, H., and Lea, A. J. (1940) An inherited jaw anomaly in long haired Dachshunds. *J. Genet.*, **39**: 285–296.
Haupt, K. H., Prieur, D. J., Moore, M. P., Hegreberg, G. A., Gavin, P. P., and Johnson, R. S. (1985) Familial canine dermatomyositis: clinical, electrodiagnostic and genetic studies. *Amer. J. Vet. Res.*, **49**: 1861–1869.
Hall, D. E. (1972) *Blood Coagulation and Its Disorders in the Dog.* Balliere Tindall.
Hall, R. L. (1978) Variability and speciation in Canids and Hominids. In Hall, R. L. and Sharp, H. S. (Editors) *Wolf and Man.* Academic Press.
Hare, W. C. D., and Bovee, K. (1974) Chromosome translocation in Miniature Poodles. *Vet. Rec.*, **95**: 218–219.
Hartley, W. J., Barker, J. S. F., Wanner, R. A., and Farrow, B. R. H. (1978) Inherited cerebellar degeneration in the Rough Coated Collie. *Aust. Vet. Pract.*, **8**: 79–85.
Hayes, H. M. (1974) Congenital umbilical and unguinal hernias in cattle, horses, swine, dogs and cats. *Amer. J. Vet. Res.* **35**: 839–842.
Healy, P. J., Farrow, B. R. H., Nicholas, F. W., Hedberg, K., Ratcliffe, R. (1984) Canine: a biochemical and genetic investigation. *Res. Vet. Sci.*, **36**: 354–359.
Hedhammar, A., Olssen, S. E., Andersson, S. A., Persson, L., Pettersson, L., Olausson, A., and Sundgren, P. E. (1979) Canine hip dysplasia. *J. Amer. Vet. Med. Assoc.*, **174**: 1012–1016.
Hegreberg, G. A., Padgett, G. A., Gorham, J. R., and Henson, J. B. (1969) A connective tissue disease of dogs. *J. Hered.*, **60**: 249–254.

Henricson, B., Norberg, I., and Olsson, S. E. (1966) On the etiology and pathogenesis of hip dysplasia: a comparative review. *J. Small Anim. Pract.*, **7**: 673–688.
Herrmann, A., and Wegner, W. (1988) Augenveranderungen bei alteren tigerteckein. *Prakt. Tierarzt*, **69**(3): 33–36.
Hickman, J., and Spickett, S. G. (1965) Avascular necrosis of the femoral head in the dog. *Proc. Roy. Soc. Med.*, **58**: 366–369.
Hime, J. M., and Drake, J. C. (1965) Osteochondrosis of the spine in the Foxhound. *Vet. Rec.*, **77**: 455–459.
Hippel, E. (1930) Embryologische untersuchungen uber vererbung angehorener karactkt uber schichstar des hundes sowie uber besondere form von kapselkatakt. *Graefes Arch. Ophthalmol.*, **124**: 300–324.
Hirschfeld, W. K. (1956) Fokkerij op genotype. *Genen Phaenen*, **1**(1): 1–5.
Hirth, R. S., and Nielsen, S. W. (1967) Familial canine globoid cell leukodystrophy. *J. Small Anim. Pract.*, **8**: 569–575.
Hodgeman, S. F. J. (1963) Abnormalities and defects in pedigree dogs. *J. Small Anim. Pract.*, **4**: 447–456.
Hollis, C., and Whitney. D. D. (1957) Jet black and off-black poodles. *Popular Dogs*. August 1957.
Humphrey, E. S., and Warner, L. (1934) *Working Dogs*. Oxford University Press.
Hutt, F. B. (1964) *Animal Genetics*. Ronald Press.
Hutt, F. B. (1967) Genetic selection to reduce the incidence of hip dysplasia in dogs. *J. Amer. Vet. Med. Assoc.*, **151**: 1041–1048.
Hutt, F. B. (1979) *Genetics for Dog Breeders*. Freeman.
Hutt, F. B., and Lahunta, A. (1971) A lethal glossopharyngeal defect in the dog. *J. Hered.*, **62**: 291–293.
Iljin, N. A. (1932) [*Genetics and Breeding of Dogs*] Selkolhozghiz.
Iljin, N. A. (1941) Wolf-dog genetics. *J. Genet.*, **42**: 359–414.
Inada, S., Sakamoto, H., Haruta, K. (1978) A clinical study on hereditary progressive neurogenic muscular atrophy in Pointer dogs. *Jap. J. Vet. Med.*, **40**: 539–547.
Inada, S., Yamauchi, C., Igata, A., Osame, S., and Izumo, S. (1986) Canine storage disease characterized by hereditary neurogenic muscular atrophy. *Amer. J. Vet. Res.*, **47**: 2294–2299.
Jaggy, A., Gaillard, C., Lang, J., and Vandevelde, M. (1988) Hereditary cervical spondylopathy (wobbler syndrome) in the Borzoi dog. *J. Amer. Anim. Hosp. Assoc.*, **24**: 453–458.
Jansen, B., Tryhonas, L., Wong, J., Thorner, P., Maxie, M. G., Valli, V. E., Baumal, R., and Basrur, P. K. (1986) Mode of inheritance of Samoyed hereditary glomerulopathy. *J. Lab. Clin. Med.*, **107**: 551–555.
Jeffreys, A. J., and Morton, D. M. (1987) DNA fingerprints of dogs and cats. *Anim. Genet.*, **18**: 1–15.
Jezyk, P. F., Haskins, M. E., MacKay-Smith, and Patterson, D. F. (1986) *J. Amer. Vet. Med. Assoc.*, **188**: 833–839.
Johnson, G. F., Gilbertson, S. R., Goldfischer, S., Grushoff, P. S., and Sternlieb, I. (1984) Cytochemical detection of inherited copper toxicosis in Bedlington terriers. *Vet. Pathol.*, **21**: 57–60.
Johnson, G. F., Sternlieb, I., Tweld, D. C., Grushoff, P. S., and Scheinberg, I. H. (1980) Inheritance of copper toxicosis in Bedlington Terriers. *Amer. J. Vet. Res.*, **41**: 1865–1866.
Johnson, G. F., Zawie, D. A., Gilbertson, S. R. and Steinlieb, I. (1982) Chronic active hepatitis in Dobermann Pinschers. *J. Amer. Vet. Med. Assoc.*, **180**: 1439–1442.
Johnson, G. S., Turrentine, M. A., and Kraus, K. H. (1988) Canine von Willebrand's Disease. *Vet. Clin. N. Amer.*, **18**: 195–229.
Johnstone, I. B., and Lotz, F. (1979) An inherited platelet function defect in Basset hounds. *Canad. Vet. J.*, **20**: 211–215.
Jones, B. R. (1984) Hereditary myotonia in the Chow Chow. *Vet. Annual*, **24**: 286–291.
Juneja, R. K., Arnold, I. C. J., Gahne, B., and Bouw, J. (1987) Parentage testing of dogs using variants of blood proteins: description of five new plasma protein polymorphisms. *Anim. Genet.* **18**: 297–310.
Karbe, E., and Schiefer, B. (1967) Familial amaurotic idiocy in male German short-haired pointers. *Pathol. Vet.*, **4**: 223–233.

Keeler, C. E., and Timble, H. C. (1938) The inheritance of dew claws in the dog. *J. Hered.*, **29**: 145–148.
Kelly, D. F. (1986) Familial renal disease in the Wheaten Terrier. *Vet. Annual*, **26**: 305–311.
Kelley, R. B. (1949) *Sheep Dogs*. Angus & Robertson.
Klein, E., Steinberg, S. A., Weiss, S. R. B., Matthews, D. M., and Uhde, T. W. (1988) Relationship between genetic deafness and fear-related behaviour in nervous Pointer dogs. *Physiol. Behav.*, **43**: 307–312.
Klinckmann, G., Koniszewski, G., and Wegner, W. (1989) Light microscopic investigations on the retina of dogs carrying the merle factor. *J. Vet. Med. Assoc.*, **33**: 674–688.
Klinckmann, G., Koniszewski, G., and Wegner, W. (1987) Lichtmikropische untersuchungen an den corneae von merle dachsunden. Dtsch. *Tierarztl. Wschr.*, **94**: 338–341.
Klinckmann, G., and Wegner, W. (1987) Tonometrien bei merlehunden. *Dtsch. Tierarztl. Wschr.*, **94:** 337–338.
Koch, S. A. (1972) Cataracts in interrelated Old English Sheepdogs. *J. Amer. Vet. Med. Assoc.*, **160**: 299–301.
Koppang, N. (1970) Neuronal ceroid-lipofuscinosis in English setters. *J. Small Anim. Pract.*, **10**: 639–644.
Kramer, J. W., Hegreberg, G. A., Bryan, G. C., Meyers, K., and Ott, R. L. (1976) A muscle disorder of labrador retrievers characterized by a deficiency of type II muscle fibres. *J. Amer. Vet. Med. Assoc.*, **169**: 817–820.
Kramer, J. W., Hegreberg, G. A., and Hamilton, M. J. (1981) Inheritance of a neuromuscular disorder of Labrador Retriever Dogs. *J. Amer. Vet. Med. Assoc.*, **179**: 380–381.
Kramer, J. W., Klaasen, J. K., Baskin, D. G., Prieur, D. J., Rantanen, N. W., Robinette, J. D., Graber, W. R., and Rashti, L. (1988) Inheritance of diabetes mellitus in Keeshond dogs. *Amer. J. Vet. Res.*, **49**: 428–431.
Kramer, J. W., Nottingham, S., Robinette, J., Lenz, G., Sylvester, S., and Dessouki, M. I. (1980) Inherited, early onset, insulin requiring diabetes mellitus of Keeshond dogs. *Diabetes*, **29**: 558–565.
Kramer, J. W., Schiffer, S. P., Sande, R. D., Rantanen, N. W., and Whitener, E. K. (1982) Characterization of heritable thoractic hemivertebra of the German Shorthaired Pointer. *J. Amer. Vet. Med. Assoc.*, **181**: 814–815.
Krhyzanowski, J., Malinowski, E., and Studnicki. W. (1975) [Pregnancy period in dog breeds] *Med. Wet.*, **31**: 373–374.
Kunkle, G. A., Chrisman, C. L., Gross, T. L., Fadok, V., and Werner, L. L. (1985) Dermatomyositis in Collie dogs. *Comp. Cont. Educ. Pract. Vet.*, **7**: 185–192.
Ladds, P. W., Dennis, S. M., and Leipold, H. W. (1971) Lethal congenital edema in Bulldog pups. *J. Amer. Vet. Med. Assoc.*, **159**: 81–86.
Ladrat, J., Blin, P. C., and Lauvergne, J. J. (1969) Ectromelie bithoracique hereditaire chez le chien. *Ann. Genet. Select. Anim.*, **1**: 119–130.
Lahunta, A., and Averill, D. R. (1976) Hereditary cerebellar cortical and extrapyramidal nuclear abiotrophy in Kerry blue terriers. *J. Amer. Vet. Med. Assoc.*, **168**: 1119–1124.
Lahunter, A., Fenner, W. R., Indrieri, R. J., Mellick, P. W., Gardner, S., and Bell, J. S. (1980) Hereditary cerebellar cortical abiotrophy in the Gordon Setter. *J. Amer. Vet. Med. Assoc.*, **177**: 538–541.
Langebaek, R. (1986) Variation in hair and skin texture in blue dogs. *Nord. Vet. Med.*, **38**: 383–387.
Laratta, L. J., Riis, R. C., Kern, T. J., and Koch, S. A. (1985) Multiple congenital ocular defects in the Akita dog. *Cornell Vet.*, **75**: 381–192.
Larsen, R. E., Dias, E., and Cervenka, J. (1978) Centric fusion of chromosomes in a bitch. *Amer. J. Vet. Res.*, **39**: 861–864.
Larsen, R. E., Dias, E., Flores, G., and Selden, J. R. (1979) Breeding studies reveal segregation of a canine Robertsonian translocation along Mendelian proportions. *Cytogenet. Cell Genet.*, **24**: 95–101.
Lau, R. E. (1977) Inherited premature closure of the distal ulnar physis. *J. Amer. Anim. Hosp. Assoc.*, **13**: 609–612.
Lavelle, R. B. (1984) Inherited dwarfism in English Pointers. *Austr. Vet. J.*, **61**: 268.
Leighton, E. A., Linn, J. M., Wilham, R. L., and Castleberry, M. W. (1977) Genetic study of canine hip dysplasia. *Amer. J. Vet. Res.*, **38**: 241–244.

Leon, A., Curtis, R., and Barnett, K. C. (1986) Hereditary persistent hyperplastic primary vitreous in the Staffordshire Bull terrier. *J. Amer. Anim. Hosp. Assoc.*, **22**: 765–774.
Lerner, I. M. (1958) *The Genetic Basis of Selection*. Chapman & Hall.
Letard, E. (1930) Experiences sur heredite mendelienne du charactere peau nue dans l'espece chien. *Rev. Vet. J. Med. Vet.*, **82**: 553–570.
Little, C. C. (1957) *The Inheritance of Coat Colour in Dogs*. Cornell Univ. Press.
Lium, B., and Moe, L. (1985) Hereditary multifocal renal cystadenocarcinomas and nodular dermatofibrosis in the German Shepherd dog. *Vet. Pathol.*, **22**: 447–455.
Loeffler, K., and Meyer, M. (1961) Erbliche Paterllarluxation bei Toy Spaniels. *Dtsche. Tierarzl. Woch.*, **68**: 619–622.
Lorenz, M. D., Cork, L. C., Griffin, J. W., Adams, R. J., and Price, D. L., (1979) Hereditary spinal muscular atrophy in Brittany Spaniels. *J. Amer. Vet. Med. Assoc.*, **175**: 833–839.
Lund, J. E., Padgett, G. A., and Gorham, J. R. (1970) Inheritance of cyclic neutropenia in the dog. *J. Hered.*, **61**: 46–49.
Lush, J. L. (1945) *Animal Breeding Plans*. Iowa State College Press.
Lust, G., and Farrell, P. W. (1977) Hip dysplasia in dogs. *Cornell Vet.*, **67**: 447–466.
Lyngset, A., and Lyngset, O. (1970) [Litter size in the dog]. *Nord. Vet. Med.*, **22**: 186–191.
MacIntosh, N. W. G. (1975) The origin of the dingo: an enigma. In Fox, M. W. (Editor) *The Wild Canids*. Van Nostrand Reinhold Co.
MacKenzie, S. A. (1985) Canine hip dysplasia. *Canine Pract.*, **12**: 19–22.
MacMillan, A. D., and Lipton, D. E. (1978) Heritability of multifocal retina dysplasia in American Cocker Spaniels. *J. Amer. Vet. Med. Assoc.*, **172**: 568–572.
Mann, G. E., and Stratton, J. (1966) Dermoid sinus in the Rhodesian Ridgeback. *J. Small Anim. Pract.*, **7**: 631–642.
Martin, C. L., and Leach, R. (1970) Everted membrane nictitans in German short-haired pointers. *J. Amer. Vet. Med. Assoc.*, **157**: 1229–1232.
Martin, C. L., and Leipold, H. W. (1974) Aphakia and multiple ocular defects in St. Bernard puppies. *Vet. Med. Small Anim.*, **69**: 448–453.
Mason, T. A. (1977) Cervical vertebral instability (wobbler syndrome) in the Dobermann. *Austr. Vet. J.*, **53**: 440–445.
Matsas, D. J., and Yang, T. J. (1980) Karyotype analysis of leucocytes of cyclic neutropenia/normal bone marrow transplant chimeras six years after transplantation. *Amer. J. Vet. Res.*, **41**: 1863–1864.
Mayr, B., Krutzler, J., Schleger, W., and Auer, H. (1986) A new type of Robertsonian translocation in the dog. *J. Hered.*, **77**: 127.
McKerrell, R. E., Anderson, J. R., Herrtage, M. E., Littlewood, J. D., Palmer, A. C. (1984) Generalized muscular weakness in the Labrador Retriever. *Vet. Rec.*, *115*: 276.
McMillan, R. B. (1970) Early canid burial from the Western Ozark highland. *Science*, **167**: 1246–1247.
Mengel, R. M. (1971) A study of dog-coyote hybrids. *J. Mammal.*, **52**: 316–336.
Merton, D. A. (1982) Known and suspected genetic diseases. *Comp. Cont. Educ. Pract. Vet.*, **4**: 332–356.
Meyers, K. M., Padgett, G. A., and Dickson, W. M. (1970) The genetic basis of a kinetic disorder of Scottish Terrier dogs. *J. Hered.*, **61**: 189–192.
Meyers, V. N., Jezek, P. F., Aquirre, G. D., and Patterson, D. F. (1983) Short-limb dwarfism and ocular defects in the Samoyed dog. *J. Amer. Vet. Med. Assoc.*, **183**: 975–979.
Meyers-Wallen, V. N., and Patterson, D. F. (1988) XX sex reversal in the American cocker spaniel dog. *Hum. Genet.*, **80**: 23–30.
Migaki, G. (1982) Compendium of inherited metabolic diseases in animals. In Desnick, R. J. (Editor) *Animal Models of Inherited Metabolic Diseases*. Alan R. Liss Inc.
Miller, L. M., Hegreberg, G. A., Prier, D. J., and Hamilton, M. J. (1984) Inheritance of congenital myasthenia gravis in smooth Fox Terrier dogs. *J. Hered.*, **75**: 163–166.
Mitchell, A. L. (1935) Dominant dilution and other colour factors in Collie dogs. *J. Hered.*, **26**: 424–430.
Moore, P. F. (1984) Systemic histiocytosis of Bernese Mountain dogs. *Vet. Pathol.*, **21**: 554–563.
Morgan, J. P., Ljunggren, G., and Read, R. (1967) Spondylosis deformans in the dog. *J. Small Anim.*, **8**: 57–66.

Narfstrom, K. (1981) Cataract in the West Highland White Terrier. *J. Small Anim. Pract.*, **22**: 467–471.
Nash, A. S., Kelly, D. F., and Gaskell, C. J. (1984) Progressive renal disease in soft coated wheaten terriers. *J. Small Anim. Pract.*, **25**: 479–487.
Nicholas, F. (1978) Pituitary dwarfism in German Shepherd Dogs. *J. Small Anim. Pract.*, **19**: 167–174.
O'Brien, J. A., and Hendriks, J. (1986) Inherited laryngeal paralysis. *Vet. Quart.*, **8**: 301–302.
Olsen, S. J., and Olsen, J. W. (1977) The Chinese wolf: ancestor of New World dogs. *Science.*, **197**: 533–535.
O'Neill, C. S. (1981) Hereditary skin disease in the dog and cat. *Comp. Cont. Educ. Pract. Vet.*, **3**: 791–800.
Osborne, C. A., Clifford, D. H., and Hessen, C. (1967) Hereditary esophageal achalasia in dogs. *J. Amer. Vet. Med. Assoc.*, **151**: 572–581.
O'Sullivan, N., and Robinson, R. (1989) Harlequin colour in the Great Dane dog. *Genetica*, **78**: 215–218.
Padgett, G. A., and Mostovsky, U. V. (1986) The mode of inheritance of craniomandibular osteopathy in West Highland White Terrier dogs. *Amer. J. Med. Genet.*, **25**: 9–13.
Palmer, A. C., and Wallace, M. E. (1967) Deformation of cervical vertebrae in Basset hounds. *Vet. Rec.*, **80**: 430–433.
Palmer, A. C., Payne, J. E., and Wallace, M. E. (1973) Hereditary quadriplegia and amblyopia in the Irish Setter. *J. Small Anim. Pract.*, **14**: 343–352.
Parry, B. W. (1985) Evaluation of haemostatic disorders in dogs and cats. *Vet. Annual*, **25**: 302–317.
Patel, V., Koppang, N., Patel, B., and Zeman, W. (1974) Phenylenediamine-mediated peroxidase deficiency in English Setters with neuronal ceroid-lipofuscinosis. *Lab. Invest.*, **30**: 366–368.
Patterson, D. F. (1974) Pathologic and genetic studies of congenital heart disease in the dog. *Advance Cardiol.*, **13**: 210–249.
Patterson, D. F. (1976) Congenital defects of the cardiovascular system of dogs. *Advanc. Vet. Sci. Comp. Med.*, **20**: 1–37.
Patterson, D. F. (1980) A catalog of genetic disorders of the dog. In Kirk, R. W. (Editor) *Current Veterinary Therapy. VII. Small Animal Practice.* W. H. Saunders.
Patterson, D. F., Haskins, M. E., and Jezyk, P. F. (1982) Models of human genetic disease in domestic animals. *Advan. Hum. Genet.* **12**: 263–339.
Patterson, D. F., Medway, W., Luginbuhl, H., and Chacko, S. (1976) Congenital hereditary lymphoedema in the dog. *J. Med. Genet.*, **4**: 145–152.
Pearson, K., and Usher, C. H. (1929) Albinism in dogs. *Biometrika*, **21**: 144–163.
Peters, J. A. (1969) Canine breed ancestry. *J. Amer. Vet. Med. Assoc.*, **155**: 621–624.
Pfaffenberger, C. J. (1963) *The New Knowledge of Dog Behaviour.* Howell Book House Inc.
Phillips, J. M. (1945) Pig jaw in the cocker spaniel. *J. Hered.*, **36**: 177–181.
Phillips, J. M. and Felton, T. M. (1934) Hereditary umbilical hernia in dogs. *J. Hered.*, **30**: 433–435.
Pick, J. R., Goyer, R. A., Graham, J. B., and Renwick, J. H. (1967) Subluxation of the carpus in dogs. *Lab. Invest.*, **17**: 243–248.
Pidduck, H. (1987) A review of inherited disease in the dog. *Vet. Annual*, **27**: 293–311.
Pidduck, H., and Webbon, P. M. (1978) The genetic control of Perthes' disease in toy poodles—a working hypothesis. *J. Small Anim. Pract.*, **19**: 729–733.
Portman-Graham, R. (1975) *The Mating and Whelping of Dogs.* Popular Dogs.
Prasse, K. W., Crouser, D., Beutler, E., Walker, M., and Schall, W. D. (1975) Pyruvate kinase deficiency anaemia with terminal myelofibrosis and osteoseclerosis in a Beagle. *J. Amer. Vet. Med. Assoc.*, **166**: 1170–1175.
Priester, W. A. (1972a) Sex, size and breed as risk factors in canine patellar luxation. *J. Amer. Vet. Med. Assoc.*, **160**: 740–742.
Priester, W. A. (1972b) Congenital ocular defects in cattle, horses, cats and dogs. *J. Amer. Vet. Med. Assoc.*, **160**: 1504–1511.
Priester, W. A. (1974) Canine progressive retinal atrophy: occurrence by age, breed and sex. *Amer. J. Vet. Res.*, **35**: 571–574.
Priester, W. A. (1976) Canine invertebral disk disease. *Theriogenology*, **6**: 293–303.

Priester, W. A. (1979) Occurrence of mammary neoplasms in bitches in relation to breed, age, tumour type, and geographical region. *J. Small Anim. Pract.*, **20**: 1–11.
Priester, W. A., and Mulvihill, J. J. (1972) Canine hip dysplasia: relative risk by sex, size and breed. *J. Amer. Vet. Med. Assoc.*, **160**: 735–739.
Pullig, T. (1950) Inheritance of whorls in Cocker Spaniels. *J. Hered.*, **41**: 97–99.
Pullig, T. (1952) Inheritance of a skull defect in Cocker Spaniels. *J. Hered.*, **43**: 97–99.
Pullig, T. (1953) Anury in Cocker Spaniels. *J. Hered.*, **44**: 105–107.
Pullig, T. (1957) Brachyury in Cocker Spaniels. *J. Hered.*, **48**: 75–76.
Rafiquzzaman, M., Svenkerud, R., Strande, A., and Hauge, J. G. (1976) Glycogenosis in the dog. *Acta Vet. Scand.*, **17**: 196–209.
Rand, J. S., Best, S. J., and Mathews, K. A. (1988) Portosystemic vascular shunts in a family of American Cocker Spaniels. *J. Amer. Anim. Hosp. Assoc.*, **24**: 265–272.
Randolph, J. F., Center, S. A., Kallsfelz, F. A. et al. (1986) Familial nonspherocytic hemolytic anaemia. *Amer. J. Vet. Res.*, **47**: 687–695.
Rapaport, F. T., and Bachvaroff, R. J. (1978) Experimental transplantation and histocompatibility systems in the canine species. *Advanc. Vet. Sci. Comp. Med.*, **22**: 195–219.
Rasmussen, P. G. (1972) Multiple epiphyseal dysplasia in Beagle puppies. *Acta Radiol. Suppl.*, **319**: 251–254.
Read, D. H., Harrington, D. D., Keenan, T. W., and Hinsman, E. J. (1976) Neuronal visceral GM1 gangliosidosis in a dog with β-galactosidase deficiency. *Science*, **194**: 442–445.
Reetz, I., Stecker, M., and Wegner, W. (1977) Audiometrische befunde in einer merlezucht. *Dtsche Tierarztl. Wschr.*, **84**: 273–277.
Rehfeld, C. E. (1970) Definition of relationships in a closed Beagle colony. *Amer. J. Vet. Res.*, **31**: 723–732.
Renshaw, H. W., and Davus, W. C. (1979) Canine granulocytopathy syndrome. *Amer. J. Pathol.*, **95**: 731–744.
Riser, W. H., Haskins, M. E., Jesyk, P. F., and Patterson, D. F. (1980) Pseudoachondroplastic dysplasia in Miniature Poodles. *J. Amer. Vet. Med. Assoc.*, **176**: 335–341.
Roberts, S. R., and Helper, L. C. (1972) Cataracts in Afghan hounds. *J. Amer. Vet. Med. Assoc.*, **16**: 427–432.
Robinson, R. (1973) Relationship between litter size and weight of dam in the dog. *Vet. Rec.*, **92**: 221–223.
Robinson, R. (1985) Chinese crested dog. *J. Hered.*, **76**: 217–218.
Robinson, R. (1988) Inheritance of colour and coat in the Belgian shepherd dog. *Genetica*, **76**: 139–142.
Robinson, R. (1989a) Inheritance of coat colour in the Anatolian shepherd dog. (in press).
Robinson, R. (1989b) Inheritance of coat colour in the Hovawart dog. *Genetica*, **78**: 121–123.
Robinson, R. (1990) Reproduction in the Hovawart breed of dog. (in press).
Rubin, L. F. (1963) Hereditary retina detatchment in Bedlington Terriers. *Small Anim. Clin.*, **3**: 387–389.
Rubin, L. F. (1968) Heredity of retina dysplasia in Bedlington Terriers. *J. Amer. Vet. Med. Assoc.*, **152**: 260–262.
Rubin, L. F. (1974) Cataract in Golden Retrievers. *J. Amer. Vet. Med. Assoc.*, **165**: 457–458.
Rubin, L. F., Bourns, T. K. R., and Lord, L. H. (1967) Hereditary of hemeralopia in Alaskan Malamutes. *Amer. J. Vet. Res.*, **28**: 355–357.
Rubin, L. F., and Flowers, R. D. (1972) Inherited cataract in a family of Standard Poodles. *J. Amer. Vet. Med. Assoc.*, **161**: 207–208.
Rubin, L. F., Koch, S. A., and Huber, R. J. (1969) Hereditary cataracts in miniature Schauzers. *J. Amer. Vet. Med. Assoc.*, **154**: 1456–1458.
Sanda, A., and Krizenecky, J. (1965) Genetic basis of necrosis of digits in shortcoated Setters. *Sb. Vys. Sk. Zemed. Brne. Rada.*, **B.13**: 281–286.
Sande, R. D., Alexander, J. E., Spencer, G. R., Padgett, G. A., and Davis, W. C. (1982) Dwarfism in Alaskan Malamutes. *Amer. J. Pathol.*, **106**: 224–236.
Sandefeldt, E., Cummings, J. F., Lahunta, A., Bjorck, G., and Krook, L. (1973) Hereditary neuronal abiotrophy in the Swedish Lapland dog. *Cornell Vet.*, **63**: 1–71.
Schaible, R. H. (1976) Linkage of a pigmentary trait with a high level of uric acid excretion in the Dalmatian dog. *Genetics*, s68.
Schaible, R. H. (1981) A Dalmatian study. *Amer. Kennel Club Gaz.*, **98** (4): 48–52.

Schaible, R. H., and Brumbangh, J. A. (1976) Electron microscopy of pigment cells in variegated and nonvariegated piebald spotted dogs. *Pigment Cell.*, **3**: 191–220.

Scott, J. P., and Fuller, J. L. (1965) *Genetics and the Social Behaviour of the Dog.* University of Chicago Press.

Scott, J. P., Fuller, J. L., and King, J. A. (1959) Inheritance of annual breeding cycles in hybrid Basenji-Cocker Spaniel dogs. *J. Hered.*, **50**: 255–261.

Selden, J. R., Moorhead, P. S., Koo, G. C., Wachtel, S. S., Haskins, M. E., and Patterson, D. F. (1984) Inherited XX sex reversal in the Cocker Spaniel dog. *Hum. Genet.*, **67**: 62–68.

Selden, J. R., Wachtel, S. S., Koo, G. C., Haskins, M. E., and Patterson, D. F. (1978) Genetic basis of *XX* true hermaphroditism: evidence in the dog. *Science*, **201**: 644–646.

Selmanowitz, V. J., Kramer, K. M., and Orentreich, N. (1970) Congenital ectodermal defect in Miniature Poodles. *J. Hered.*, **61**: 196–199.

Selmanowitz, V. J., Markofsky, J., and Orentreich, N. (1977a) Heritability of an ectodermal defect. *J. Dermatol. Surg. Oncol.*, **3**: 623–626.

Selmanowitz, V. J., Markofsky, J., and Orentreich, N. (1977b) Black hair follicular dysplasia in dogs. *J. Amer. Vet. Med. Assoc.*, **171**: 1079–1081.

Shive, R. J., Hare, W. C. D., and Patterson, D. F., (1965) Chromosome anomalies in dogs with congenital heart disease. *Cytogenetics*, **4**: 340–348.

Smith, J. E., Moore, K., Arens, M., Rinderknecht, G. A., and Ledet, A. (1983) Hereditary elliptocytosis with protein band 4.1 deficiency in the dog. *Blood*, **61**: 373–377.

Sponenberg, D. P., (1985) Inheritance of the harlequin color in Great Dane dogs. *J. Hered.*, **76**: 224–225.

Sponenberg, D. P., and Bowling, A. T. (1985) Heritable syndrome of skeleton defects in a family of Australian Shepherd dogs. *J. Hered.*, **75**: 393–394.

Sponenberg, D. P., and Lahunter, A. (1981) Hereditary hypertropic neuropathy in Tibetan Mastiff dogs. *J. Hered.*, **72**: 287.

Sponenberg, D. P., and Lamoreux, M. L. (1985) Inheritance of tweed, a modification of merle, in the Australian Shepherd dogs. *J. Hered.*, **79**: 303–304.

Sponenberg, D. P., Scott, E., and Scott, W. (1988) American Hairless Terriers: a recessive gene causing hairlessness in dogs. *J. Hered.*, **79**: 69.

Spurling, N. W. (1980) Hereditary disorders of haemostasis in dogs. *Vet. Bull.*, **50**: 151–173.

Stabenfeldt, G. H., and Shille, V. M. (1972) Reproduction in the dog and cat. In Cole, H. H., and Cupps, P. T. (Editors) *Reproduction in Domestic Animals.* Academic Press.

Stades, F. C. (1978) Retinal dysplasia in the dog. *Tijdschr. Diergeneesk.*, **103**: 1087–1090.

Steinberg, H. S., Troncoso, J. C., Cork, L. C., and Price, D. L. (1981) Clinical features of inherited cerebellar degeneration in Gordon Setters. *J. Amer. Vet. Med. Assoc.*, **179**: 886–890.

Stockard, C. R. (1936) An hereditary lethal for localized motor and preganglionic neurones, with resulting paralysis, in the dog. *Amer. J. Anat.*, **59**: 1–53.

Stockard, C. R. (1941) The genetic and endocrine basis for differences in form and behaviour in dogs. *Amer. Anat. Mem.*, **19**: 775Pp.

Taylor, R. M., and Farrow, B. R. H. (1988) Fucosidosis in English Springer Spaniel dogs. *Comp. Pathol. Bull.*, **20**: 2–4.

Taylor, R. M., Martin, I. C. A., and Farrow, B. R. H. (1989) Reproductive abnormalities in canine fuscosidosis. *J. Comp. Pathol.*, **100**: 369–380.

Tedor, J. B., and Reif, J. S. (1978) Natal patterns among registered dogs in the United States. *J. Amer. Vet. Med. Assoc.*, **172**: 1179–1185.

Thornbrug, L. P., Shaw, D., Dolan, M., Raisbeck, M., Crawford, S., Dennis, G. L., and Olwin, D. B. (1986) Hereditary copper toxicosis in West Highland White Terriers. *Vet. Path.*, **23**: 148–154.

Tjebbes, K., and Wriedt, C. (1927) The albino factor in the Samoyed dog. *Hereditas*, **10**: 165–168.

Treu, H., Reetz, I., Wegner, W., and Krause, D. (1976) Andrologische befunde in einer merlezucht. *Zuchthyg.*, **11**: 49–61.

Trimble, H. D., and Keeler, C. E. (1938) The inheritance of high uric acid excretion in dogs. *J. Hered.*, **19**: 280–289.

Troncoso, J. C., Cork, L. C., and Price, D. L. (1985) Canine inherited ataxia. *J. Neuropathol. Exper. neurol.*, **44**: 165–175.

Turba, E., and Willer, S. (1987) Untersuchungen zur Vererbung von hasen scharten und wolfsrachen beim Deutschen Boxen. *Mh. Vet. Med. 42:* 897-901
Valentine, B. A., Cooper, B. J., Cummings, J. F., and Lahunter, A. (1986) Progressive muscular dystrophy in a Golden Retriever dog. *Acta Neuropathol.*, **71**: 301–310.
Valentine, B. A., Cooper, B. J., Dietze, A. E., and Noden, D. M. (1988) Canine congenital diaphragmatic hernia. *J. Vet. Intern. Med.*, **2**: 109–112.
Venker-Haagen, A. J., Bouw, J., and Hartman, W. (1981) Hereditary transmission of laryngeal paralysis in Bouviers. *J. Amer. Anim. Hosp. Assoc.*, **17**: 75–76.
Venker-Haagen, A. J., Hartman, W., and Goedegeuure, S. A. (1978) Spontaneous laryngeal paralysis in young Bouviers. *J. Amer. Vet. Med. Assoc.*, **14**: 714–720.
Vriesendorp, H. M. (1979) Applications of transplantation immunology in the dog. *Advanc. Vet. Sci. Comp. Med.*, **23**: 229–265.
Vriesendorp, H. M., Hartog, B., Smid-Mered, B. M. J., and Weslbroek, D. L. (1974) Immunogenetic markers in canine paternity cases. *J. Small Anim. Pract.*, **15**: 693–699.
Wallace, M. E. (1975) Keeshonds: a genetic study of epilepsy and EEG readings. *J. Small Anim. Pract.*, **16**: 1–10.
Wallace, M. E., and Palmer, A. C. (1984) Recessive mode of inheritance in myasthenia gravis in the Jack Russell Terrier. *Vet. Rec.*, **114**: 350.
Wallin, B., (1986) Perthes sjukdom hos west higland white terrier, en gebetisk studie. *Svensk. Vet.*, **38**: 114–118.
Walvroot, H. C., (1985) Glycogen storage disease type II in the Lapland dog. *Vet. Quart.*, **7**: 187–190.
Walvroot, H. C., Koster, J. F., and Reuser, A. J. J. (1985a) Heterozygous detection in a family of Lapland dogs with a recessively inherited disease: canine glycogen storage disease type II. *Res. Vet. Sci.*, **38**: 174–178.
Walvroot, H. C., Nes, J. J., Stokhof, A. A., and Wolvekamp, W. T. C. (1984a) *J. Amer. Anim. Hosp. Assoc.*, **20**: 279–286.
Walvroot, H. C., Slee, R. G., Sluis, K. J., Koster, J. F., and Reuser, A. J. J. (1984b) Biochemical genetics of the Lapland dog: model of the glycogen storage disease type II. *Amer. J. Med. Genet.*, **19**: 589–598.
Waring, G. O., MacMillan, A., and Reveles, P. (1986) Inheritance of crystalline corneal dystrophy in Siberian Huskies. *J. Amer. Anim. Hosp. Assoc.*, **22**: 655–658.
Watson, A. D. J., and Dixon, R. T. (1977) Cystic bone lesions in related Old English Sheepdogs. *J. Small Anim. Pract.*, **18**: 561–571.
Weber, W. (1959) Ueber die vererbung der medain nasenspalte beim hund. *Schweiz. Arch. Tierheilk.*, **101**: 378–381.
Weber, W., and Freudiger, U. (1977) Erbanalytiske untersuchungen uber die chronische exocrine pankreasinsuffizienz beim Deuchschen Schsferhunde. *Schwiez. Arch. Tierheilk.*, **119**: 257–263.
Wentinck, G. H., Hartman, W., and Doeman, J. P. (1974) Three cases of myotonia in a family of chows. *Tijdschr. Diergeneesk.*, **99**: 729–731.
Wentick, C. H., Linde-Sipman, J. S., Meijer, A. E. F. H., Kamphuise, H. A. C., Vorstenbosch, C. J. A. H. V., Hartman, W., and Hendricks, H. J. (1972) Myopathy with a possible recessive *X* linked inheritance in a litter of Irish Terriers. *Vet. Pathol.*, **9**: 328–349.
Westermarck, E. (1980) Hereditary nature of canine pancreatic degenerative atrophy in the German Shepherd dog. *Acta Vet. Scand.*, **21**: 389–394.
Whitbresd, T. J., Gill, J. J. B. Gill, and Lewis, D. G. (1983) An inherited enchondrodystrophy in the English Pointer dog. *J. Small Anim. Pract.*, **24**: 399–461.
White, K. (1978) *Dogs: Their Mating, Whelping and Weaning.* K. & K. Books.
Whitney, D. D. (1952) Silver Poodles. *Popular Dogs.* August 1952.
Whitney, D. D. (1958) Black and Silver Poodles. *Popular Dogs.* August 1958.
Whitney, L. F. (1947) *How to Breed Dogs.* Orange Judd Publishing Co.
Wilcock, B. P., and Patterson, J. M. (1979) Familial glomerulonephritis in Doberman Pinscher dogs. *Canad. Vet. J.*, **20**: 244–249.
Willeberg. P. K. W., Kastrap, W., and Andresen, E. (1975) Pituitary dwarfism in German Shepherd dogs: studies on somatomedin activity. *Nord. Vet. Med.*, **27**: 448–454.
Williamson, J. H., (1979) Intersexuality in a family of Kerry Blue Terriers. *J. Hered.*, **70**: 138–139.

Willis, M. B. (1976) *The German Shepherd Dog.* K. & R Books.
Willis, M. B. (1989) *The Genetics of the Dog.* Witherby.
Willis, M. B., Barnett, K. C., and Tempest, W. M. (1979) Genetic aspects of lens luxation in the Tibetan terrier. *Vet. Rec.*, **104**: 409–412.
Winge, O. (1950) *Inheritance in Dogs.* Constable.
Wolfe, E. D., Vainisi, S. J., and Santos-Anderson, R. (1978). Rod-cone dysplasia in the Collie. *J. Amer. Vet. Med. Assoc.*, **173**: 1331–1333.
Wolf, E. D., Vainisi, S. J., and Santos-Anderson, R. (1979) Rod-cone dysplasia and retina atrophy. *J. Amer. Vet. Med. Assoc.*, **174**: 324,337.
Workman, M. J., and Robinson, R. (1989) Coat colours of the Cavalier King Charles Spaniel. (in press).
Wouda, W., and Nes, J. J. (1986) Progressive ataxia due to central demyelination in Rottwieler dogs. *Vet. Quart.*, **8**: 89–97.
Wreidt, C. (1925) Letale faktoren. *Z. Tierz. Zuchbiol.*, **3**: 223–230.
Yakely, W. L. (1972) Collie eye anomaly: decreased prevalence through selected breeding. *J. Amer. Vet. Med. Assoc.*, **161**: 1103–1107.
Yakely, W. L. (1978) A study of heritability of cataracts in the American Cocker Spaniel. *J. Amer. Vet. Med. Assoc.*, **172**: 814–817.
Yakely, W. L., Hedreburg, G. A., and Padgett, G. A. (1971) Familial cataracts in the American Cocker Spaniel. *J. Amer. Anim. Hosp. Assoc.*, **7**: 127–135.
Yakely, W. L., Wyman, M., Donovan, E. F., and Fechheimer, N. S. (1968) Genetic transmission of an ocular fundus anomaly in Collies. *J. Amer. Vet. Med. Assoc.*, **152**: 457– 461.
Yang, T. J. (1978) Recovery of coat color in cyclic neutropenia/normal bone marrow transplants. *Amer. J. Pathol.*, **91**: 149–154.
Yentzen, Y. (1965) Color inheritance in German Shepherd dogs. *German Shep. Dog Rev.*, **43 (3)**: 14–22.
Zeuner, F. E. (1963) *A History of Domesticated Animals.* Hutchinson.
Zulueta, A. (1949) The hairless dogs of Madrid. *Proc. VIII Congr. Genet.*, 687–688.

Index

The entries may be used as a glossary for genetic terms by referring to the cited page(s).

LIBRARY
WITHDRAWN
WARWICKSHIRE COLLEGE